AF903970

# COMPOSITE FILAMENT WINDING

**Edited by S.T. Peters**

ASM International®
Materials Park, Ohio 44073-0002
www.asminternational.org

Copyright © 2011
by
ASM International®
All rights reserved

No part of this book may be reproduced, stored in a retrieval system, or transmitted, in any form or by any means, electronic, mechanical, photocopying, recording, or otherwise, without the written permission of the copyright owner.

First printing, September 2011
Second printing, June 2024

Great care is taken in the compilation and production of this book, but it should be made clear that NO WARRANTIES, EXPRESS OR IMPLIED, INCLUDING, WITHOUT LIMITATION, WARRANTIES OF MERCHANTABILITY OR FITNESS FOR A PARTICULAR PURPOSE, ARE GIVEN IN CONNECTION WITH THIS PUBLICATION. Although this information is believed to be accurate by ASM, ASM cannot guarantee that favorable results will be obtained from the use of this publication alone. This publication is intended for use by persons having technical skill, at their sole discretion and risk. Since the conditions of product or material use are outside of ASM's control, ASM assumes no liability or obligation in connection with any use of this information. No claim of any kind, whether as to products or information in this publication, and whether or not based on negligence, shall be greater in amount than the purchase price of this product or publication in respect of which damages are claimed. THE REMEDY HEREBY PROVIDED SHALL BE THE EXCLUSIVE AND SOLE REMEDY OF BUYER, AND IN NO EVENT SHALL EITHER PARTY BE LIABLE FOR SPECIAL, INDIRECT OR CONSEQUENTIAL DAMAGES WHETHER OR NOT CAUSED BY OR RESULTING FROM THE NEGLIGENCE OF SUCH PARTY. As with any material, evaluation of the material under end-use conditions prior to specification is essential. Therefore, specific testing under actual conditions is recommended.

Nothing contained in this book shall be construed as a grant of any right of manufacture, sale, use, or reproduction, in connection with any method, process, apparatus, product, composition, or system, whether or not covered by letters patent, copyright, or trademark, and nothing contained in this book shall be construed as a defense against any alleged infringement of letters patent, copyright, or trademark, or as a defense against liability for such infringement.

Comments, criticisms, and suggestions are invited, and should be forwarded to ASM International.

*Prepared under the direction of the ASM International Technical Book Committee (2010–2011), Michael J. Pfeifer, Chair.*

*ASM International staff who worked on this project include Scott Henry, Senior Manager, Content Development and Publishing; Eileen De Guire, Senior Content Developer; Victoria Burt, Content Developer; Sue Sellers, Editorial Assistant; Bonnie Sanders, Manager of Production; Madrid Tramble, Senior Production Coordinator; and Diane Whitelaw, Production Coordinator.*

Library of Congress Control Number: 2011908767

ISBN-13: 978-1-61503-722-3 (print)
ISBN-10: 0-61503-722-5 (print)
ISBN: 978-1-61503-852-7 (pdf)
ISBN: 978-1-62708-338-6 (electronic)

SAN: 204-7586

ASM International®
Materials Park, OH 44073-0002
www.asminternational.org

Printed in the United States of America

Copyright © 2011, ASM International®
All rights reserved.
www.asminternational.org

# Contents

Copyright © 2011, ASM International®
All rights reserved.
www.asminternational.org

# Preface

THIS IS THE FIRST CONTEMPORARY BOOK to treat filament winding with a focus on automation. The purpose of this book is to provide detailed procedures for automated filament winding for complex shapes, to chronicle the progress that has been made in robotic filament winding along with ancillary processes that support the deposition process. It is also to show what has and what can be done to advance the process and science of low-cost composite fabrication by filament winding. The need for this book has been evident because the bulk of filament winding machinery purchases now have a greater number of control axes and substantial automation in ancillary operations, such as mandrel removal, and cure (Chapter 2). There has also been a geographical shift with the process. Literature searches have shown that more professional papers now emerge from Canada, South Korea, and the Netherlands than from the United States. Investigators at the Korea Advanced Institute of Science and Technology have developed two new iterations of the standard test and evaluation bottle (Chapter 8).

This book builds on three previous publications, *Filament Winding, Composite Structure Fabrication*, SAMPE Publishers, 1991 and 1999; and, S. Koussios, *Filament Winding, A Unified Approach*, Delft University Press, 2004; along with the many individual publications and presentations of the authors of the present work. The previous books addressed mainly aerospace applications and offered hands-on instructions for making components using the rudimentary, by today's standards, winders. Among other areas, the previous books presented geodetic and nongeodetic roving trajectories, kinematics with collision control, and derived articulated pressurizable structures.

Four of the nine chapters in this book (Chapters 2 to 5) address automation directly. Now the highest automation is used in the commercial arena, for example, pipes. The kinds of structures that can be produced are shown in Chapters 6 and 7. Finally, Chapter 9 addresses the actual processes of pressure vessel analysis, fabrication, and testing.

I want to dedicate this book to two mentors and friends, Yuri Tarnopol'skii and Lowrie McLarty, who both gently prodded me toward composite physical and mental activity and who corrected some of my more glaring errors. I still hold them in my memory. I would also like to acknowledge the excellent editorial staff at ASM International. They were thoroughly competent, professional, and furthermore, easy to work with.

Stan T. Peters

Copyright © 2011, ASM International®
All rights reserved.
www.asminternational.org

# About the Editor

Mr. Stan T. Peters is General Manager of Process Research, an engineering consulting firm in Mountain View, Calif. He has been editor of the *SAMPE Journal* and remains a member of the editorial review board for the *Journal of Advanced Materials.*

He has written or edited two books, *Filament Winding: Composite Structure Fabrication*, 1991, 1999; and *Handbook of Composites*, 2nd ed., Chapman and Hall, 1998; and has 32 publications and 3 patents.

He was elected a SAMPE Fellow in 1990, was a Westinghouse Fellow Engineer, and received the Westinghouse Signature Award in 1990, the WEC ESG Special Achievement Award in 1992, and the Boeing Certificate of Achievement for the EELV/Delta IV program in May 2004. He is a professional engineer, manufacturing, State of California, and has taught or presented tutorials on composites in a number of venues.

Composite Filament Winding
S.T. Peters, editor

Copyright © 2011, ASM International®
All rights reserved.
www.asminternational.org

# CHAPTER 1

# Filament Winding— Introduction and Overview

S.T. Peters, Process Research, Consultants

## Introduction

The objective of this book is the facilitation of fabrication of useful composite shapes by the most efficient and least costly method: filament winding. A previous book on filament winding, *Filament Winding; Composite Structure Fabrication* (Ref 1.1), was published in 1994 by the Society for the Advancement of Material and Process Engineering. It was aimed at the large audience of filament winders who were using two- and three-axis machines with either mechanical or computer controls. The audience has changed. Now, most winders have computer controls and at least three axes. Winding with four axes is common because the shapes of the products have evolved to include more complexity. However, the automation used on the winding machine and ancillary components does not eliminate the need for proper fiber handling. This book is a primer on how to use the new machines, starting with an overview of what is available for automation, a look at what has been done, and a primer on handling the fibers with minimum damage prior to laydown.

## Computer and Computer Controls

There has been great progress in machine automation during the last 16 years. Much of the progress can be attributed to the impact of computers, although much of this work has been accomplished outside of the United States. For example, most publications in the realm of reinforced piping have originated in Canada, and a significant number of papers on rocket motors/pressure vessels have had South Korea as their source.

The advances in machine control give the winders an opportunity to wind noncylindrical and nonsymmetric objects and allow for a much more precise laydown of fiber. The higher capacities of computer memory that is now available allow the control of up to six axes of motion: mandrel rotation, cross feed, horizontal carriage movement, vertical carriage movement, wind eye rotation, and wind eye yaw (Fig. 1.1).

Control of ancillary components is now common so that there can be automated mandrel placement, automatic tie-on and cut-off, automated loading of wet fiber-covered mandrels into the curing oven, and finally, automated extraction. The new machine systems have allowed pipe manufacturers the ability to wind rather than hand lay-up, spray-up, resin transfer mold, or compression mold their elbows, reducers, and tees. Filament winding these components results in better mechanical properties through the use of continuous fibers, with the accompanying lowering of manufacturing costs (Ref 1.2).

## Composite Analysis

Increasingly better methods for composite analysis have given designers a quicker, more reliable answer for closed-form analysis. There are "B"-basis allowable data for fiber/resin combinations for 121 °C (250 °F)-curing prepreg for the aircraft composite component industry (e.g., AGATE) (Ref 1.3), but there is no comparable

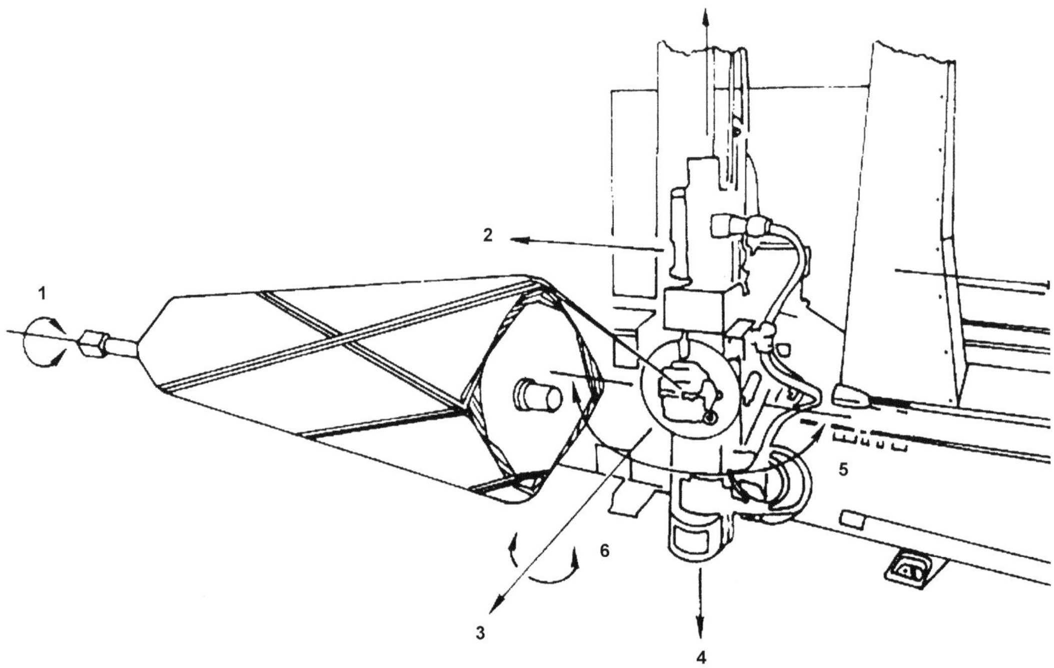


**Fig. 1.1** Schematic illustration of the manufacture of a tapered box beam on a computer-controlled six-axis filament winder. Courtesy of Mclean Anderson. Source: Ref 1.1, © S.T. Peters, 2009

database for filament-wound composites that use prepreg tow and wet resin-impregnated tow or roving. The principles of composite analysis and "B"-basis allowables are discussed more extensively in Chapters 8 and 7, respectively.

## Structural Analysis

Structural analysis has also progressed so that structures can be modeled and the model transferred into any of several finite-element analysis programs, which can also incorporate laminate analysis routines and libraries of composite micromechanical properties for fiber/resin combinations. These tools can accomplish in one day what normally took several months in 1994, when Ref 1.1 was published.

## Pressure Vessels

Pressure vessel designers can benefit from the analysis from netting, closed-form, and finite-element analysis routines incorporated in the Composite Design and Analysis Code. There are several case dome configurations shown in Chapter 8. Design has also benefited from the wider use of the Standard Test and Evaluation Bottle, enabling dome and skirt testing and optimization as well as providing information on the pressure vessel. These subjects are more fully elaborated on in Chapter 8.

Two new designs for small pressure vessels have gained wider use. Isotensoid vessels (Fig. 1.2) are essentially composite pressure vessels with Isotensoid domes but without a cylindrical section. They are proposed for use as liquid propane gas containers and would be placed in the automobile spare wheel well. For automotive use, rigorous tests must be imposed. Additional information is presented in Chapter 8.

Toroidal pressure vessels (Fig. 1.3) are in commercial use to contain air for breathing. Advantages quoted for the QinetiQ, United-Kingdom-manufactured doughnut-shaped composites, are lighter weight, better use of volume, and greater protection for the pressure regulator (Ref 1.4, 1.5).

## Pipes

Pipe manufacturers have seen that multiangle filament-wound pipe can confer some desirable properties, such as addition of higher bending strength for a pipe simply supported at the ends, hoop layers to improve buckling strength for

Fig. 1.2 Composite Isotensoid pressure vessel. Courtesy of Advanced Lightweight Engineering B.V.

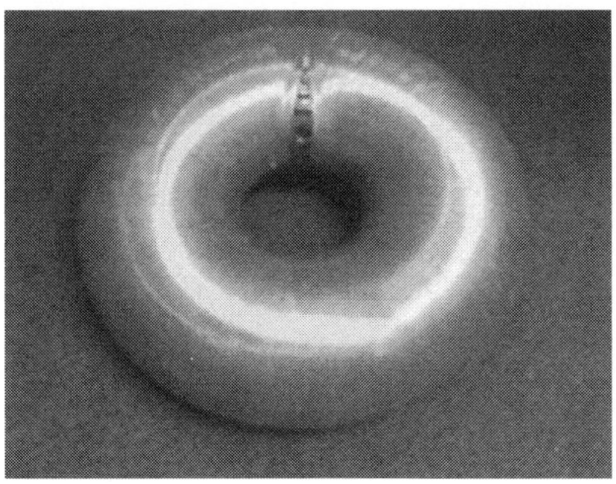

Fig. 1.3 Toroidal pressure vessel. Courtesy of M. Hinton and QinetiQ. Source: Ref 1.5

buried pipe, or more axial strength with low-angle helicals for a vertical pipe (Ref 1.6).

Fig. 1.4 Proton-M launch vehicle. Interstage just above strapons. Courtesy of Khrunichev State Research and Production Space Center

## Grid Structures

The commercialization of grid structures has been realized on the Proton-M launch vehicle, which uses a composite Anisogrid structure for interstage (first to second stage is open, and second to third stage is overwound with aramid/epoxy), payload attachment fittings, and composite fairing (Fig. 1.4). A 15% weight savings is reported for the lattice adapter compared to other adapters (Ref 1.7, 1.8).

The Isotruss, developed at Brigham Young University, is a three-dimensional configuration that can be filament wound but, in the more complicated designs, is more aptly fabricated with a technology that looks more like weaving.

One of the initial Isotruss applications was a composite bicycle frame weighing less than 3 lb (1.36 kg) (Fig. 1.5) (Ref 1.9).

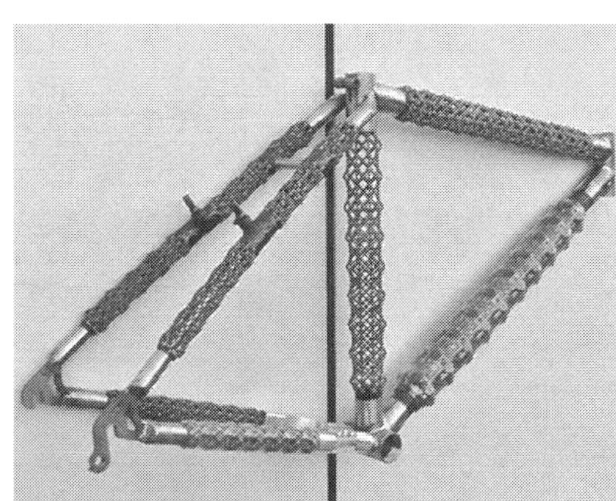

Fig. 1.5 Isotruss bicycle frame

## Deep Sea Oil Platform Drill Risers

Composite drill and production riser development began almost 30 years ago, based on perceptions that they would reduce deck loads and provide several other benefits. The National Institute of Standards and Technology was the funding agent to initiate these programs. The principal composite contractors are Spencer Composites Corp. (Lincoln, CA), which is working with Kvaerner (Oslo, Norway), and Northrop Grumman (Sunnyvale, CA), which is working with ABB Vetco Gray (Houston, TX). The key technologies are the metal-to-composite interface (MCI) and the liner. The Spencer MCI uses a proprietary traplock design, a 0.12 in. (3 mm) titanium liner, and titanium flanges, which make it interchangeable with the standard titanium drilling riser currently used on the Heidrun tension leg platform in Norway. The riser has been in service since 2001. The Northrop Grumman MCI is the convex version of the wound-in-place hyperboloid joint invented at Westinghouse for the lower joint of the Tomahawk Missile Launcher. The Northrop Grumman riser has been slated for sea trial by Petrobras, Brazil since 2003 but has not seen service yet. Some problems remain; for example, the composite drill riser is reported to be approximately 30% to 3 times the cost of the steel riser, and the composite wall is thicker, which may negatively influence hydraulics (Ref 1.10–1.16).

## High-Speed Rotors

The design of high-speed rotors now benefits from the use of very high-strength carbon fibers such as Toray T 800 and T 1000 (Toray Composites America, Tacoma WA). The research into high-speed rotor design started with filament-wound monolithic rings made with conventional epoxies, then liquid polyurethane elastomers were introduced; continuous curing was effected, and finally, multiring composite rotors have been selected for further study. Each iteration had a positive effect on the upper limit of rotational speed (Ref 1.17–1.19).

## Filament-Wound Preforms

The filament-winding process is a low-cost method of manufacturing composite preforms, which have a high fiber volume with a virtually unlimited number of repeatable and accurate fiber angles. As an alternate to woven fabric and braiding, Storage Tek Composites (Louisville, CO; now part of Oracle, Inc.) used dry high-strength carbon fiber, a prepreg delivery head, and a polyester scrim cloth that is melted at the ends of the mandrel to hold the fibers in place for resin transfer molding (Ref 1.20).

### REFERENCES


1.1 S.T. Peters, W.D. Humphrey, and R.F. Foral, *Filament Winding; Composite Structure Fabrication*, 2nd ed., SAMPE Publishers, Covina, CA, 1994, ©S.T. Peters, 2009

1.2 J. Gilchrist, Automated Filament Winding of Large, Complex Shapes, *SAMPE International Symposium*, May 2000 (Long Beach, CA), p 1452

1.3 AGATE, E-765 Approved Databases, Fibercote Industries, Waterbury, CT

1.4 J. Cook and J. Chambers, Toroidal Pressure Vessels for Breathing Apparatus, *SAMPE Paris Conference*, April 22–24, 1998, p 125–138

1.5 S. Li and J. Cook, An Analysis of Filament Overwound Toroidal Pressure Vessels and Optimum Design of Such Structures, *J. Pressure Vessel Technol.*, Vol 124, May 2002, p 215–222

1.6 R. Lea and C. Yang, "Improving the Mechanical Properties of Composite Pipe Using Multi-Angle Filament Winding," Paper 458, NACE Corrosion '98

1.7 Y. Bakhvalov, V. Molochev, S. Petrokovskii, V. Barynin, V. Vasiliev, and A. Razin, "Proton-M Composite Interstage Structures: Design, Manufacturing and Performance," European Conference for Aerospace Sciences (EUCASS), July 2005 (Moscow)

1.8 V. Vasiliev, A. Razin, G. Totaro, and F. De Nicola, "Aniosgrid Conical Adapters for Commercial Space Applications," AIAA paper 2005-3440, p 2099–2107

1.9 D. Jensen, A Glimpse into the World of Innovative Composite Isotruss Grid Structures, *SAMPE J.*, Vol 36 (No. 5), Sept–Oct, 2000, p 8–16

1.10 J. Price, The "State of the Art" in Composite Material Development and Applications for the Oil and Gas Industry,

*Proceedings of the 12th (2002) International Offshore and Polar Engineering Conference* (Kitakyushu, Japan), 2002, p 125–130
1.11 M. Jellison, R.B. Chandler, G. Prideco, M. Payne, and J. Shepard, "Drillstring Technology Vanguard for World-Class Extended-Reach Drilling," OTC 18512, 2007 Offshore Technology Conference, April–May 2007 (Houston, TX)
1.12 D. Johnson, Demand for Composites in Deepwater Oil Production, *SAMPE International Symposium*, May 2000 (Long Beach, CA), p 2105–2117
1.13 Current Composite Drilling Riser Projects, *Compos. World*, Jan. 2, 2002
1.14 G. Galle and J. Whitehead, A Successful Program for Composites in the Oil and Gas Industry, *SAMPE International Symposium*, May 2000 (Long Beach, CA), p 2128–2135
1.15 D. Johnson, K. Lo, and H. Wu, Development of Rigid Composite Risers—A Status Report, *SAMPE International Symposium*, May 2000 (Long Beach, CA), p 2136–2198
1.16 B. Spencer, personal communication, March 4, 2010
1.17 C. Gabrys and C. Bakis, Design and Manufacturing of Filament Wound Elastomeric Matrix Composite Flywheels, *J. Reinf. Plast. Compos.*, Vol 16 (No. 6), 1997, p 488–502
1.18 C. Gabrys and C. Bakis, Filament Winding of Thick Composite Rings with High Radial Accretion Rates, *Proceedings of the International Exposition*, The Society of the Plastics Industry, New York, 1998, p 1904–1915
1.19 S. Ha, H. Kim, and S. Han, Effects of Rotor Sizes and Epoxy System on the Process-Induced Residual Strains within Multi-Ring Composite Rotors, *J. Compos. Mater.*, Vol 38 (No. 10), 2004, p 871–884
1.20 D. Howell and F. Roundy, Filament Wound Preforms for RTM, *SAMPE J.*, Vol 36 (No. 2), March–April 2000, p 41–46

Composite Filament Winding
S.T. Peters, editor

Copyright © 2011, ASM International®
All rights reserved.
www.asminternational.org

# CHAPTER 2

# Automated Filament Winding Systems

John E. Green, The Green Sales Guy, Inc. (GSG, Inc.)

## Introduction

This chapter reviews the progress of filament winding and the automated filament winding process found in today's state-of-the-art filament winding facilities. The present-day filament winding process has been in use for over 50 years, but there has been little modernization during the first 40 of those years. Modernization has generally been confined to the past decade or so.

Filament winding machines have used simple low-tech practices for many years. These included manually loading mandrels, manually mixing resin and adding it to the resin bath, manually tying on the fibers to the mandrel to start winding, manually changing between different types of fibers in the same winding process, manually cutting the rovings at the conclusion of the winding process, and manually unloading the wet-wound mandrels into the oven, and so on.

Finally, there has been a revolution in updating the filament winding process. This includes fully automated winding facilities where the mandrels are loaded/unloaded with a robot or handling device. Automatic resin-mixing systems with an on-demand level gage mounted in the resin bath can discharge a specified amount of premixed resin directly into the resin bath. Fully automatic tying on of wet or prepreg rovings to the mandrel was invented along with automatic cutting off of wet or prepreg rovings and reconnecting these wet or prepreg rovings onto the next mandrel. Multiple-fiber delivery systems can be used interchangeably during winding to provide two or more types of fibers to a single winding pattern. Automatic loading of the wound parts into a curing oven and so on has also been incorporated in the modern filament winding process.

## The Beginnings of Modern Filament Winding

Richard E. Young began what could be considered modern-day filament winding in the early 1940s (Fig. 2.1). He started with a simple mechanical winder that laid fiberglass onto a wooden mandrel. After the winding process was completed, a paintbrush was used to apply an epoxy resin to the outside of the wound unit. The wound part was then inserted into an oven for curing. After curing, the wooden mandrel was removed to complete the part fabrication. Young's filament winding process was the first to be presented to the U.S. government for rocket technology applications.

From this time in the early 1940s until well into the 1980s, most filament winding machines were simple mechanically controlled pieces of equipment that used a gear box to determine the winding angle and a chain of a given length to define the length of the part to be wound. Figure 2.2 shows an early Entec vertical winding machine used to build a large rocket motor. To make a change to the winding angle, one had to manually adjust the gearbox each and every time a wind angle change was desired. This required understanding the relationship that the sprockets had within these gearboxes and how these related to winding angles that were produced on the manufactured parts. Also, each time a part length was changed, the operator had to shorten or lengthen the chain that moved the carriage by counting chain sections (links) to obtain the precise winding length desired.

**Fig. 2.1** Richard E. Young, the founder of modern-day filament winding

**Fig. 2.2** 1960s vertical winder at Thiokol in Brigham City, Utah, USA

Needless to say, mechanical filament winding machines were not very user-friendly. The operator had to deal with sprockets and carriage chain lengths for each different part configuration. This type of mechanical winder was used well into the 1980s. If a specific part was needed, and many of the very same part, a mechanical winder was the ultimate low-cost equipment. Changing over from one part diameter of a given length to another part diameter or length was very time-consuming. It took hours to calculate sprocket and chain length conversion. In addition, a filament winding company had to have many different sprocket sizes and chain lengths in stock to make these changes.

For more than 50 years, these types of mechanical filament winding machines were standard equipment supplied by four to five commercial filament winding machine manufacturers or were "home built" by the customer.

## Filament Winding Advances in the Computer Age

The next advancement in filament winding was the photo eye winder. This winder had a large drum with a black-and-white print on the drum to allow the winder to follow a "pattern" for the winding angle. To change the winding angle meant removing the black-and-white paper on the drum and substituting a different black-and-white pattern on a different piece of paper. The winder would then follow the new pattern to generate a different winding angle. This was a marked improvement over the simple mechanical winders of the day. These photo eye winders were sold in the late 1970s and into the 1980s. A large-diameter tank winder is shown in Fig. 2.3.

The evolution of filament winding machines continued with the addition of computer control. Pattern-generation software for personal computers was developed for winding machines. Entec was one of the first companies in the industry to offer their own in-house-created computer-controlled filament winding machine coupled with filament winding pattern-generation software. This combination allowed the operator to create winding patterns that would run directly on the winder. The winding machine could now be "taught" to follow a programmed path.

As with computers in the early 1980s, the winding machines of this time were a vast improvement over the mechanical and photo eye winders they replaced. Changing the winding angle was much easier and required no manual modification of gears or chain lengths. The servo motors and electronic controllers used by the new winding machines also greatly improved the efficiency of filament winding. A McClean Anderson three-spindle winder is shown in Fig. 2.4. The new winding machines that used these computer controls were still manufacturing parts one by one with little automation. The operators would load a single mandrel at a time, hand tie-on the fibers, (automatically) wind the part, cut the fibers by hand, and then remove the wound part to a curing oven or curing rack.

**Fig. 2.3** Pipe and tank winder

**Fig. 2.4** McClean Anderson three-spindle winder

Filament-wound part production was still very manually intensive. Little thought, at this time, was given to high-volume production or automation. Figure 2.5 shows a polar winder with a circ winding arm on the left.

For larger parts, the operator of an overhead crane or forklift would load and unload the mandrel each time, and curing ovens often held only a few parts at a time. This was not conducive to high-speed production, but many parts of that era were aerospace or defense related, and high-volume production was not required.

Fig. 2.5 Polar and helical winder

The exception to this was the oil and chemical industry, which needed many parts wound at a time to increase efficiency.

## Oil and Gas Industry Production Requirements

Advancements in oil and chemical pipe production in the late 1970s led to a production system for filament winding of pipe products. This consisted of high-speed winding machines with a 360° delivery eye where hundreds of rovings could be placed onto the mandrel at the same time (Fig. 2.6). This is called a 360° ring delivery eye, and one pass of the carriage makes one complete layer of fiber coverage on the mandrel. This high-speed winder necessitated development of a curing oven with six or more stations to cure numerous parts at a time, because curing time was always much longer than the winding time.

The curing oven had a component rotation system to turn the parts during cure. The six-unit oven rotated each mandrel in the oven. Figure 2.7 shows the pipe winders. Next to the pipe winders are the high-production curing ovens. The whole system could be indexed so a cured part could be removed for postcure processing as a newly wound part was inserted for curing. With this higher-volume curing capacity, the extraction system for pipes and tubes required increased efficiency. Extractors were developed

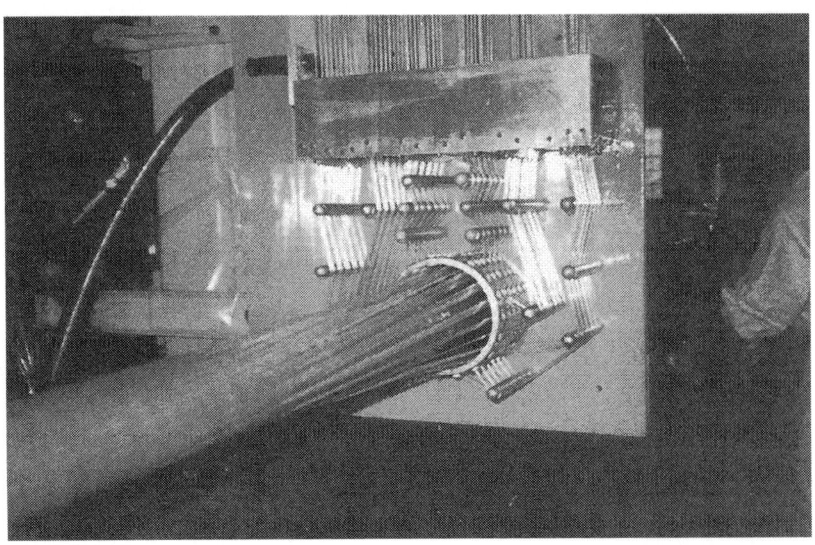

Fig. 2.6 360° ring delivery-eye system

to remove a part off the mandrel in seconds rather than minutes. High-speed extractors, complete with automated cutting/grinding of the pipe ends, were set up to handle the higher-volume production needs.

This pipe and tube manufacturing technology did not progress into other industries until the late 1980s. This is when filament-wound parts found a niche market in the sporting goods industry.

## High-Volume Production

### *Sporting Goods Manufacture*

The sporting goods industry needed large volumes of parts rapidly produced at a low cost. To fulfill this need, the filament winding industry developed multiple-spindle winders such as the 12-spindle winder in a six-by-six configuration, shown in Fig. 2.8. The multiple-spindle wind-

**Fig. 2.7** Two-spindle pipe winder with in-line curing station

**Fig. 2.8** Twelve-spindle winder

ers, such as the three-spindle winder in Fig. 2.9, allow an operator to load 10-, 20-, or even 30-spindle filament winding machines with mandrels. The winding machine could wind multiple patterns onto the part in a very short amount of time. For example, an operator could load a 20-spindle filament winding machine, tie on the fibers, wind 20 parts, and unload them in less than 10 min. Advancements in automatic applying and cutting of the fibers onto the mandrels were developed in the early 1990s for filament winding machines used in sporting goods manufacturing, but the technology was not really perfected at this time.

These multiple-spindle winders were the first big step toward automating filament winding machines and increasing efficiency of the filament winding process. One drawback, however, was that multiple-spindle winders were generally part-specific, and it was not easy to process a wide range of different part diameters and lengths on these specialized winders.

### *Propane Tank Manufacture*

In the late 1990s, the demand for lightweight, strong propane tank designs became apparent, and in response, companies developed new tank products manufactured by filament winding. To obtain the efficiency and cost-savings needed to mass produce these tanks, manufacturing companies worked closely with filament winding machine manufacturers to develop one of the most advanced, automated filament winding systems built to date. Propane tank manufacturing companies wanted to wind hundreds of thousands of components per year to meet their production goals, and few, if any, winders in the world, up to this point, had ever produced

**Fig. 2.9** Three-spindle, four-axes winder to produce pressure vessels

such large quantities on a single filament winding machine

The filament winding machine eventually designed to meet this production target was a single five-by-five (ten-spindle) winder with four axes of motion. The axes of motion are spindle rotation, horizontal carriage, radial motion, and rotating eye. Full automation was developed to make this winder highly efficient. Parts could be wound 24 hours a day, 7 days a week, and 365 days a year.

The propane (liquefied petroleum gas, or LPG) tank manufacturing process begins by creating the tank liners by blow or rotational molding (Fig. 2.10). The fibers used for winding the tank bodies are produced in spools of center-pull fiberglass. Full pallets of fiberglass spools are brought to the winder, ready to run through the resin bath and fiber-delivery system. The initial fiber stringing is done manually. When the fibers are initially strung in place, they run through a fully automated, two-part thermoset resin mixing system consisting of two resin baths side by side. At the beginning of the winding process, the fibers pass through a dip-type resin bath (Fig. 2.11). The resin is supplied on demand while being premixed in a separate mixing system. Due to the large volume of resin needed, the resin is supplied to the facility in rail tankers. The winder system incorporates a resin-level sensing and automatic filling system to bring more resin to the bath as needed. As noted previously, there are two resin baths. This redundant resin bath is required should the primary resin system begin to "go off" or gel, rendering it unusable for further manufacturing. If a change-out of the resin bath is needed, a built-in system lifts the many fibers from the nonfunctional resin bath. The nonfunctional resin bath then shifts to the side, a fresh resin bath moves into place, the fibers are lowered into the new resin bath, and the winding process continues. This change-out of resin baths happens in a matter of seconds. The bath with the old resin is automatically cleaned as soon as the current bath requires changing. The operator's job is then to just watch to make sure all is running smoothly.

The winder system also uses a manipulating arm that secures five mandrels at one time—in this case, the five plastic liners that will be used as the mandrels. The manipulating arm picks up five of the liners from an automated delivery belt and mounts them into the five-spindle winder at

**Fig. 2.10** Rotational molders to create the liners

**Fig. 2.11** Dip-type resin bath with fibers strung through the bath without resin

the same time (Fig. 2.12). The winder operator does not need to handle the mandrels during the operating cycle.

When the fibers are "threaded" through the resin bath, they are manually tied on to each of the five mandrels by the operator. After the initial fiber tie-on process is completed, no other manual attachment or cutting of fibers is needed. Follow-on mandrels are automatically loaded into the five-spindle winder, and the fibers are automatically tied onto the mandrels. This is done using a system known as auto tie-on/auto cut-off. When the winding process is complete, the automatic cut-off system cuts the wet fibers. Five more mandrels are mounted into the winding machine. Then, the wet fibers are automatically attached to the liners, and the winding process begins again. The operator is not required

**Fig. 2.12** Autoloading of wound parts into curing oven using a manipulating arm

**Fig. 2.13** Handling device loading five wound and cured parts into a conveyor system

to touch the wet rovings or mandrels during the normal course of tank winding.

The same manipulating arm that loads the liners for winding now loads the wet-wound tanks into a curing oven. After all five tanks are indexed into the oven, five fully cured tanks are picked up by a manipulating arm and moved onto a conveyor belt system that moves the cured parts to the quality-control area.

This automated filament winding process is very efficient and uses very little manual labor. The manipulating arm unloads five cured parts onto a conveyor system that is then sent to the quality-assurance area (Fig. 2.13). A few employees supervise the entire process. No manual manipulation of the mandrels or fibers is needed. Current production using a five-spindle winder equipped with this automated winding process can be over 500,000 parts per year on a single five-by-five spindle winder. This same type of technology is used in the high-speed production of pipes and tubes as well as compressed natural gas (CNG) cylinders. Multiple-spindle winders with two, three, or four axes of motion are used.

### *Curing Ovens*

There are various types of high-volume production curing ovens. These can be convection heated using gas or electricity (Fig. 2.14), internal mandrel oil heating (Fig. 2.15), or internal mandrel steam heating. The ovens or curing stations are generally built to handle from three to fifty wound mandrels at a time.

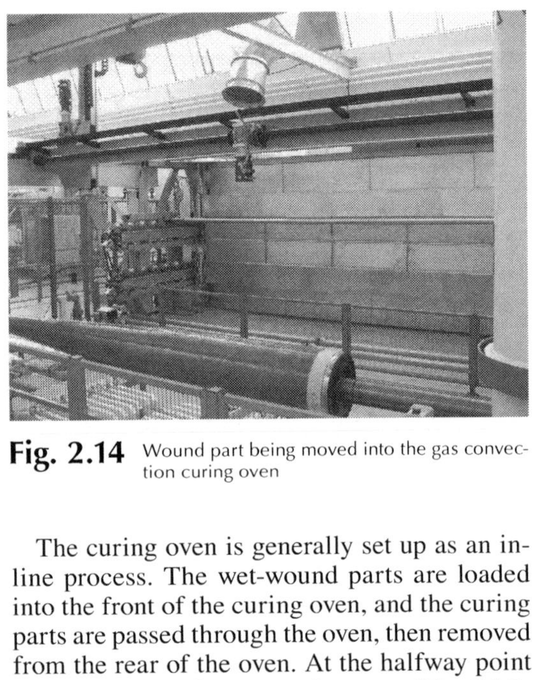

**Fig. 2.14** Wound part being moved into the gas convection curing oven

The curing oven is generally set up as an in-line process. The wet-wound parts are loaded into the front of the curing oven, and the curing parts are passed through the oven, then removed from the rear of the oven. At the halfway point through the continuous curing oven (Fig. 2.16), each part is weighed for quality control. If a part passes, it continues on to the final cure stage. If a part does not pass the quality-control weigh station, it is removed and sent back to the high-production winder for additional resin and fibers. At the rear of the oven, a manipulating arm or overhead crane unloads the cured parts onto an automatic mandrel extraction system. The extractor (Fig. 2.17, 2.18) pushes or pulls the mandrels out of the cured part and moves the

**Fig. 2.15** Three-station curing system using internal hot oil or steam for curing

**Fig. 2.16** Continuous curing oven with part rotation

**Fig. 2.17** Cured part being moved into the extractor

mandrels back to their original storage rack. If the same-sized mandrel is going to be used for more winding, it can be loaded into the lower section of the curing oven, where it will be preheated and sent back to the winding machine. The first manipulating arm or overhead crane can then pick up these preheated mandrels and load them or other sized mandrels that have been unloaded from the storage rack into the multiple-spindle winding machine, and the winding cycle begins again.

The extracted parts from Fig. 2.19 are then moved to a finishing process area, where the part ends can be sanded, tapered, or cut, depending on the application.

Again, just a few operators supervise the entire winding process. Many thousands of parts can be mass produced on such an automated filament winding system while keeping labor hours to a minimum.

**Fig. 2.18** High-speed extractor holding the mandrel and using a face plate to push off the part

**Fig. 2.19** High-speed, fully automated, 50 ton extractor

### *Filament Winding Versatility*

Similar types of automated filament winding systems are used to manufacture thermoplastic filament-wound parts. First, a manipulating arm loads multiple mandrels onto the winding machine (Fig. 2.20). This can be a "5 + 5" or comparable winding machine arrangement, with five spindles on one side of the winder and five spindles on the other side of the winder.

One side of the five-spindle winder is engaged in the winding and curing process while the other five spindles are being unloaded and the parts moved to the removal conveyor system. The winder is then loaded with new mandrels. An automated fiber tie-on system (Fig. 2.21) attaches the fibers to the mandrels. No manual attaching or cutting of the fibers is necessary.

While the parts are being wound, the thermoplastic material is being cured using a special curing system applied during the winding process. The parts are cured as they wind, and no postcure is required. After the parts are wound and cured on the winder, a manipulating arm removes them and sends the parts to postprocessing as necessary.

Three- to five-spindle winding machines are used for high-volume production of hundreds of thousands of pressure vessels each year. A two-spindle winder is shown in Fig. 2.22. If one counts up the potential production of all the

**Fig. 2.20** Five-spindle, high-production winder shown winding liquefied petroleum gas (propane) tanks

winding machines used throughout the world, there are literally millions of pressure vessels being built each and every year.

Today's automated, high-production filament winders are producing parts such as golf shafts, hockey sticks, pipes, tubes, poles, CNG tanks (Fig. 2.23), LPG tanks, self-contained breathing apparatus tanks, hydrogen tanks, electrical isolators, rollers, drive shafts, and many other items too numerous to list.

Complex-shaped parts, such as elbows and T-parts for the pipe, chemical, and oil industry, are also being mass produced on high-speed-production filament winding machines (Fig. 2.24).

Traditionally, elbows and T-shaped parts (Fig. 2.25) have been made manually or with simple two-axis winders. With a six-axis winder built specifically for complex-shaped parts, thousands of elbow and T-parts can be made automatically.

Computer control systems now fully automate filament winding systems. The computer controls the entire process, including winding machine and mandrel preheat and curing oven operation, extractor movement, operation of the manipulating arm or overhead crane, and part removal to postcure processing operations. These computer control systems are offered from Siemens (Fig. 2.26), ABB, Bosch, and many others. Also, a few of the filament winding machine companies still develop their own in-house-specific computer control systems.

Today's focus on high-production filament-wound parts includes wind turbine blades, pressure vessels, pipes, tubes, storage tanks, drive shafts, rollers, and many other parts.

Wind turbine blades, for example, can be wound using a multiple-axis, dual-delivery head

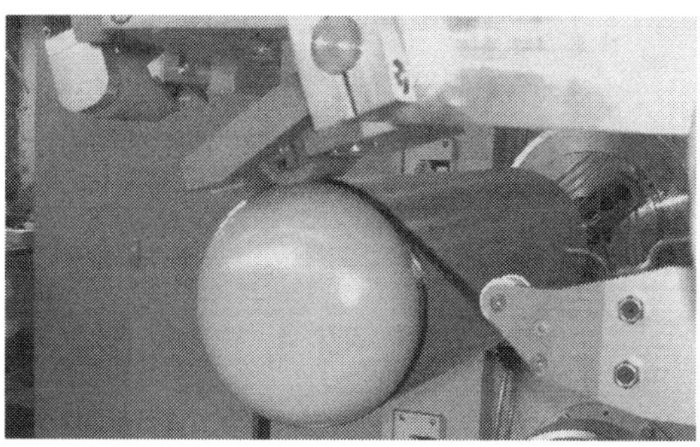

**Fig. 2.21** Automatic tie-on and cut-off system

**Fig. 2.22** Trolley cart and overhead crane automatically loading long mandrels into the two-spindle winding machine

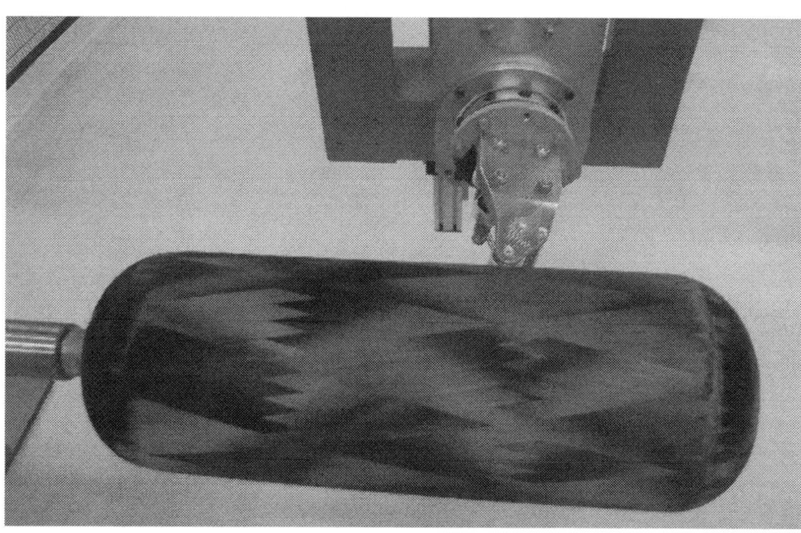

**Fig. 2.23** Filament winding a pressure vessel with carbon fiber

Fig. 2.24 Six-axes winder for making elbows and T-shaped parts

Fig. 2.25 Pipes, joints, reducers, elbows, and T-shaped parts used in the oil and chemical industry

winding machine (Fig. 2.27). This eleven-axis winder—spindle axes with two delivery system sides for winding fibers from both sides of the part simultaneously with five axes each—is used to produce wind turbine blades.

Filament winding is still one of the most cost-effective ways to produce composite products. Now, the availability of fully automated turn-key systems has eliminated the majority of the manual labor that was required as little as 20 to 40 years ago. The days of manual operations, such as loading and unloading fibers, tying and cutting off fibers, filling resin baths, and so on, are generally in the past. Today's filament winding has really become high tech and fully automated.

Fig. 2.26 Siemens Sinumeric 840D computer control system for up to 32 axes of motion

Fig. 2.27 Eleven-axes wind blade winder

## ACKNOWLEDGMENTS

The author would like to thank the following companies for use of information and/or photos in this chapter:

- CNC Technics Pvt. Ltd., Hyderabad, India
- EHA (formerly Bolenz & Schaefer, BSD), Steffenberg, Germany
- Entec Composite Machines, Salt Lake City, Utah, USA
- McClean Anderson, Scofield, Wisconsin, USA

Composite Filament Winding
S.T. Peters, editor

Copyright © 2011, ASM International®
All rights reserved.
www.asminternational.org

# CHAPTER 3

# Integral Design for Filament Winding—Materials, Winding Patterns, and Roving Dimensions for Optimal Pressure Vessels

Sotiris Koussios, Delft University of Technology

## Introduction

This chapter outlines an integral methodology for the design of cylindrical pressure vessels, with emphasis on the establishment of winding patterns and the interaction between the real fiber bed geometry (finite roving dimensions) and the theoretical one. Depending on the particular interests of the reader, this chapter can be regarded as a design manual for pressure vessels, a tutorial on winding patterns, or a collection of laminate thickness estimation techniques.

To highlight the materials-shape/pattern-roving interaction, an outline of the basic principles of pressure vessel design is inevitable. The required geometric data are combined with the roving dimensions to determine a range of feasible patterns that comply with strength (minimum required thickness) and production process demands. After a short section on laminate thickness approximation techniques (essential for establishing a range of acceptable roving dimensions), the methodology is demonstrated by an example developed from an initial set of design parameters up to the final stage, including patterns, roving dimensions, and production time minimization.

## Pressure Vessel Design

### *Shell Stresses as Determined by the Vessel Geometry*

A composite pressure vessel can be regarded as a shell of revolution, covered by a particular winding pattern (Fig. 3.1). The revolved meridian profile is given by some function, for example, $z(\rho)$ (Fig. 3.2). The loads typically consist of internal pressure, $P$, and, in the case of lifting bags, an externally applied axial load, $F_a$ (Fig. 3.3). As demonstrated in Chapter 5 regarding fiber trajectories on shells of revolution, two main shell directions can be distinguished perpendicu-

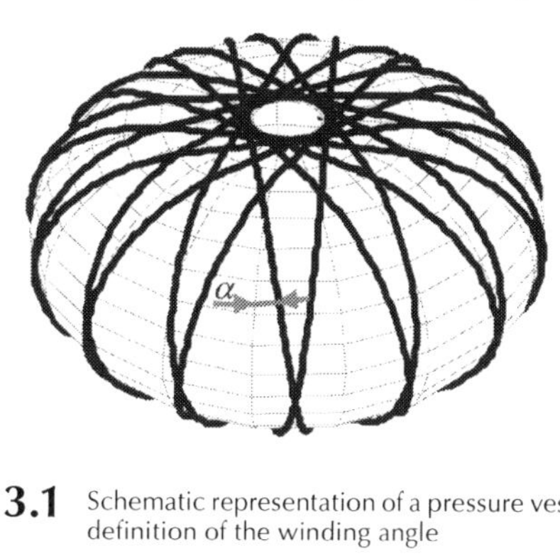


**Fig. 3.1** Schematic representation of a pressure vessel and definition of the winding angle

lar to each other (Fig. 3.2): the meridional (north ↔ south, thick black line representing $\phi$ = constant) and the parallel direction (east ↔ west, peripheries corresponding to $\rho$ = constant).

The wall thickness ($t$) of the shell is considered very small, so there are no bending effects (Ref 3.1, 3.2). Therefore, the shell stresses $\{\sigma_m, \sigma_p\}$ corresponding to the main shell directions {meridional, parallel} are located in-plane (Fig. 3.3). In addition, since every winding intersects the meridian with both $\{+\alpha, -\alpha\}$ orientations (Fig. 3.1, 3.2), there will be no in-plane shear. Hence, the shell stress vector is $\{\sigma_m, \sigma_p, 0\}$. Considering a small shell element, the normal force resultant, $F_n$, can be derived as follows (Fig. 3.3a):

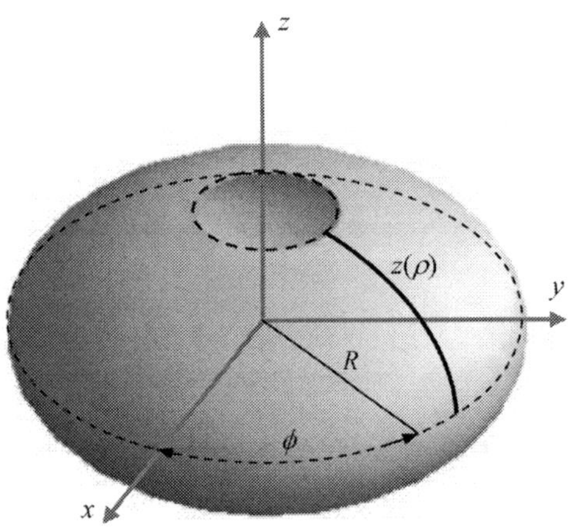


**Fig. 3.2** Geometry of a pressure vessel

$$F_n = P(R_m d\theta)(R_p d\omega) = \sigma_m t(R_p d\omega)\sin\left(\frac{d\theta}{2}\right) + \sigma_p t(R_m d\theta)\sin\left(\frac{d\omega}{2}\right) \rightarrow$$

$$\rightarrow \frac{P}{t}(R_m d\theta)(R_p d\omega) \approx \sigma_m (R_p d\omega) d\theta + \sigma_p (R_m d\theta) d\omega$$

$$\rightarrow \frac{P}{t} = \frac{\sigma_m (R_p d\omega) d\theta + \sigma_p (R_m d\theta) d\omega}{(R_m d\theta)(R_p d\omega)} = \frac{\sigma_m}{R_m} + \frac{\sigma_p}{R_p} \qquad \text{(Eq 3.1)}$$

To completely determine the unknown stresses $\{\sigma_m, \sigma_p\}$, a second condition is needed. This is the equilibrium of forces in the $z$-direction (Fig. 3.3b) (Ref 3.3, 3.4):

$$F_a + P\pi\rho^2 = \sigma_m (2\pi\rho t)\cos\beta = \sigma_m (2\pi\rho t)\frac{dz}{ds} = \sigma_m (2\pi\rho t)\frac{\rho}{R_p} \rightarrow$$

$$\rightarrow \sigma_m = \frac{R_p}{2\pi\rho^2 t}(F_a + P\pi\rho^2) = \frac{PR_p}{2t}\left(\frac{F_a}{P\pi\rho^2} + \frac{\pi\rho^2}{\pi\rho^2}\right) = \frac{PR_p}{2t}(\underline{F_a} + 1) \qquad \text{(Eq 3.2)}$$

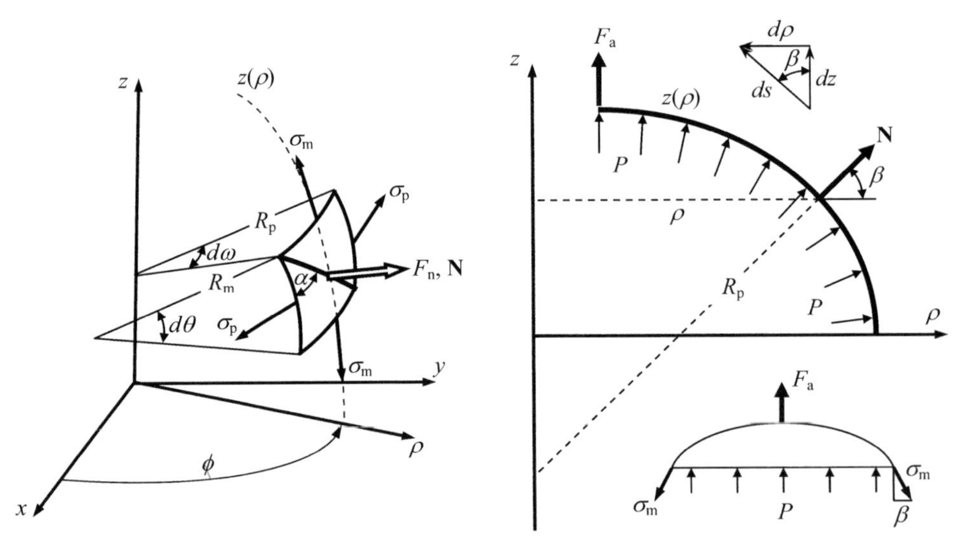


**Fig. 3.3** Equilibrium of forces and geometry for a membrane shell element with a roving under an angle $\alpha$

The combination of Eq 3.1 and 3.2 leads to:

$$\sigma_p = R_p\left(\frac{P}{t} - \frac{\sigma_m}{R_m}\right) = 2\frac{PR_p}{2t} - \frac{R_p}{R_m}\left[\frac{PR_p}{2t}\left(\underline{F_a}+1\right)\right] \rightarrow$$

$$\sigma_p = \frac{PR_p}{2t}\left[2 - \frac{R_p}{R_m}\left(\underline{F_a}+1\right)\right]$$
$$= \frac{PR_p}{2t}\left(\underline{F_a}+1\right)\left[\frac{2}{\left(\underline{F_a}+1\right)} - \frac{R_p}{R_m}\right] \rightarrow$$

$$\sigma_p = \sigma_m\left(\frac{2}{\underline{F_a}+1} - \frac{R_p}{R_m}\right) \quad \text{(Eq 3.3)}$$

Next, the following relations are introduced:

$$\rho = cY \rightarrow R = cY_{ec}$$
$$z = cZ$$
$$F_a = rP\pi R^2 \quad \text{(Eq 3.4)}$$

where $R$ is the maximum radius (at the equator of the vessel, Fig. 3.2), and $c$ is the minimum one (polar opening). The combination of Eq 3.1 and 3.2 followed by the substitution of Eq 3.4 gives:

$$\sigma_m = \frac{PR_p}{2t}\left(\frac{Y^2 + rY_{eq}^2}{Y^2}\right) \quad \text{(Eq 3.5a)}$$

$$\sigma_p = \sigma_m\left[2\left(\frac{Y^2}{Y^2 + rY_{eq}^2} - \frac{R_p}{R_m}\right)\right] \quad \text{(Eq 3.5b)}$$

In the same dimensionless parameters $\{Y, Z\}$, the radii of the meridional and parallel curvatures are (Ref 3.4–3.6):

$$R_m = -c\frac{(1+Z'(Y)^2)^{3/2}}{Z''(Y)}, \quad R_p = -c\frac{Y\sqrt{1+Z'^2(Y)}}{Z'(Y)} \quad \text{(Eq 3.6)}$$

Substitution into Eq 3.5(b) leads to:

$$\frac{\sigma_p}{\sigma_m} = Y\left[2\left(\frac{Y}{Y^2 + rY_{eq}^2}\right) - \frac{Z''(Y)}{Z'(Y) + Z'^3(Y)}\right] \quad \text{(Eq 3.7)}$$

which implies that the meridian shape does fully control the shell stress ratio.

### *Layer Stresses as Controlled by the Winding Angle*

The winding process results in a $\{+\alpha, -\alpha\}$ angleply laminate with thickness $t$. The stiffness matrix of a single layer is (Ref 3.7):

$$\mathbf{S}_0 = \begin{bmatrix} \frac{E_1}{1-\nu_{12}\nu_{21}} & \frac{\nu_{12}E_1}{1-\nu_{12}\nu_{21}} & 0 \\ \frac{\nu_{12}E_1}{1-\nu_{12}\nu_{21}} & \frac{E_2}{1-\nu_{12}\nu_{21}} & 0 \\ 0 & 0 & G_{12} \end{bmatrix} \quad \text{(Eq 3.8)}$$

where $\nu_{12}E_2 = \nu_{21}E_1$, and $E_1$ is the elasticity modulus in the fiber direction. The stiffness of the shell wall (angleply laminate) is:

$$\underline{\mathbf{S}} = \frac{1}{2}[\mathbf{S}(\alpha) + \mathbf{S}(-\alpha)]$$
$$\mathbf{S}(\alpha) = \mathbf{M}(\alpha)\mathbf{S}_0\mathbf{M}^T(\alpha) \quad \text{(Eq 3.9)}$$

in which:

$$\mathbf{M}(\alpha) = \begin{bmatrix} \cos^2\alpha & \sin^2\alpha & \sin(2\alpha) \\ \sin^2\alpha & \cos^2\alpha & -\sin(2\alpha) \\ -\cos\alpha\sin\alpha & \cos\alpha\sin\alpha & \cos(2\alpha) \end{bmatrix} \quad \text{(Eq 3.10)}$$

As a result of the shell stress vector $\underline{\sigma} = \{\sigma_m, \sigma_p, 0\}^T$, the stresses in the individual layers are:

$$\boldsymbol{\sigma}(\alpha) = \begin{Bmatrix} \sigma_1 \\ \sigma_2 \\ \tau_{12} \end{Bmatrix} = \mathbf{S}_0 \cdot \left(\mathbf{M}^T\right)^{-1} \cdot \underline{\mathbf{C}} \cdot \underline{\boldsymbol{\sigma}} \quad \text{(Eq 3.11)}$$

where $\underline{S} = \underline{C}^{-1}$. The performance of a pressure vessel is defined as the ratio of the stored volumetric energy (pressure × volume) divided by the weight of the structure. Vasiliev (Ref 3.8) and de Jong (Ref 3.3) have shown that for maximizing the performance of a pressure vessel, the shear stress $\tau_{12}$ must be zero. Setting $\tau_{12} = 0$ in Eq 3.11 results in:

$$\frac{\sigma_p}{\sigma_m} = \frac{\sin^2\alpha + k_e\cos^2\alpha}{\cos^2\alpha + k_e\sin^2\alpha} \quad \text{(Eq 3.12a)}$$

where

$$k_e = \frac{E_2\left(1+\nu_{12}\right)}{E_1\left(1+\nu_{21}\right)} \quad \text{(Eq 3.12b)}$$

For maximum strength (not necessarily performance), $k_e$ should be replaced with $k_s$, which is the ratio of the transverse to the longitudinal (fiber direction) strength of a single layer (Ref 3.8). For the netting case (matrix stiffness negligible), $k_e = 0$; for isotropic materials, $k_e$ becomes equal to 1.

It can also be shown that optimally reinforced composite pressure vessels rely on geodesic fiber trajectories (Ref 3.3, 3.8):

$$\sin\alpha = \frac{1}{Y} \qquad \text{(Eq 3.13)}$$

Plugging this in, Eq 3.12 gives:

$$\frac{\sigma_p}{\sigma_m} = \frac{1+k_e\left(Y^2-1\right)}{\left(Y^2-1\right)+k_e} \qquad \text{(Eq 3.14)}$$

For a given fiber-matrix combination ($k_e$) and geodesic winding ($Y$), this expression provides the optimal shell stress ratio.

## *Optimal Vessel Geometry*

To derive the optimal vessel shape, $Z(Y)$, the shell stresses ratio as given by Eq 3.7 and 3.14 should match:

$$\frac{\sigma_p}{\sigma_m} = Y\left[2\left(\frac{Y}{Y^2+rY_{eq}^2}\right) - \frac{Z''(Y)}{Z'(Y)+Z'^3(Y)}\right]$$
$$= \frac{1+k_e\left(Y^2-1\right)}{\left(Y^2-1\right)+k_e} \rightarrow$$

$$\int\left[2\left(\frac{Y}{Y^2+rY_{eq}^2}\right) - \frac{Z''(Y)}{Z'(Y)+Z'^3(Y)}\right]dY$$
$$= \int\frac{1}{Y}\left[\frac{1+k_e\left(Y^2-1\right)}{\left(Y^2-1\right)+k_e}\right]dY \rightarrow$$

$$\ln\left[Y^2+rY_{eq}^2\right] - \ln[Z'(Y)] + \frac{1}{2}\left[1+Z'^2(Y)\right] + \ln\left[\frac{1}{\gamma}\right]$$
$$= -\ln[Y] + \frac{1}{2}(1+k_e)\ln\left[\left(Y^2-1\right)+k_e\right] \rightarrow$$

$$\frac{Y^2+rY_{eq}^2}{\gamma}\,\frac{\sqrt{1+Z'^2(Y)}}{Z'^2(Y)} = \frac{1}{Y}\left(Y^2-1+k_e\right)^{\frac{k_e+1}{2}} \qquad \text{(Eq 3.15)}$$

where $\gamma$ is an integration constant. Solving for $Z'(Y)$ leads to:

$$Z'(Y) = \pm\frac{Y(Y^2+rY_{eq}^2)}{\sqrt{\left(k_e+Y^2-1\right)^{k_e+1}\gamma^2 - Y^2\left(Y^2+rY_{eq}^2\right)^2}} \qquad \text{(Eq 3.16)}$$

At $Y = Y_{eq}$, the slope of the meridian profile is infinite; hence, the denominator is zero (the meridian profile is vertical at that point, Fig. 3.1). Solving for the integration constant $\gamma$ results in:

$$\gamma = \frac{(1+r)Y_{eq}^3}{\left(k_e+Y_{eq}^2-1\right)^{\frac{k_e+1}{2}}} \qquad \text{(Eq 3.17)}$$

The differential equation for the meridian profile thus becomes:

$$Z'(Y) = \pm\frac{Y(Y^2+rY_{eq}^2)}{\sqrt{\left(\frac{k_e+Y^2-1}{k_e+Y_{eq}^2-1}\right)^{k_e+1}(1+r)^2Y_{eq}^6 - Y^2\left(Y^2+rY_{eq}^2\right)^2}} \qquad \text{(Eq 3.18)}$$

The denominator nullifies again when $Y$ approaches the value 1. However, according to Eq 3.13, $Y$ cannot become less than 1. To ensure real values for $\alpha$ and $Z'(Y)$, $Y_{min}$ is defined as:

$$Y_{min} = \max\left[\begin{matrix} Y \text{ solution of } \left(\frac{k_e+Y^2-1}{k_e+Y_{eq}^2-1}\right)^{k_e+1}(1+r)^2Y_{eq}^6 - Y^2\left(Y^2+rY_{eq}^2\right)^2 = 0 \\ 1 \end{matrix}\right] \qquad \text{(Eq 3.19)}$$

Note that for $k_e = 0$ (netting approach), $Y_{min}$ becomes equal to 1 only when $r = -1/q$. For $k_e > 0$, $Y_{min}$ may even become smaller than 1. In general, $Y_{min(k_e>0)}$ is closer to 1 than $Y_{min(k_e=0)}$ can reach.

With this limit, the optimal meridian profile can be obtained by integration of Eq 3.18 over the interval $[Y_{min}, Y_{eq}]$. The design procedure relies on only three parameters:

- Ratio of the equatorial radius and polar opening radius, $Y_{eq}$ (approximately, since $Y_{min}$ does not exactly become equal to 1 unless $r$ or $k_e$ is specially adjusted to that) (Ref 3.4)

- The true axial load ($F_a$), expressed as a fraction of the axial load induced by the internal pressure over the projected surface at the equator $r = F_a/(P\pi R^2)$
- Degree of material orthotropy, $k_e$ (Eq 3.12b)

Next to these parameters, one can define the dimensionless length $H = h/c$ of a (optional) cylindrical part to be added (Fig. 3.4).

Figure 3.5 provides various profiles for $r = 0$ and $Y_{eq} = 5$, based on different $k$ values. The role of the material "anisotropy" parameter (indicated now as $k$ instead of $k_e$) is clearly visible. For $k = 1$, a sphere is nearly obtained, whereas for $k = 0$, the form becomes a pure netting-based isotensoid (where all fibers are equally loaded in tension).

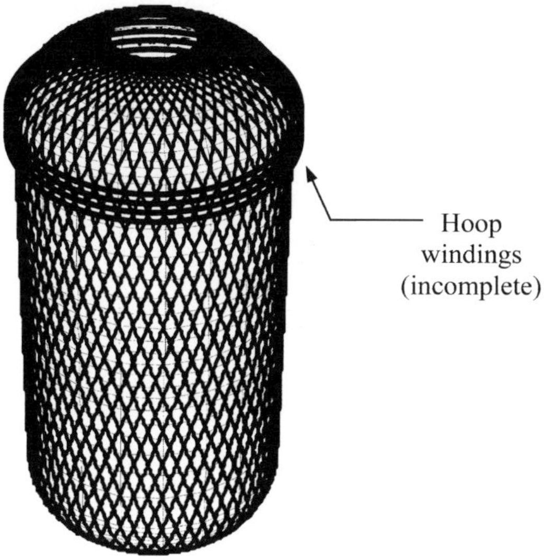


**Fig. 3.4** Polar and hoop windings on a dome-cylinder combination

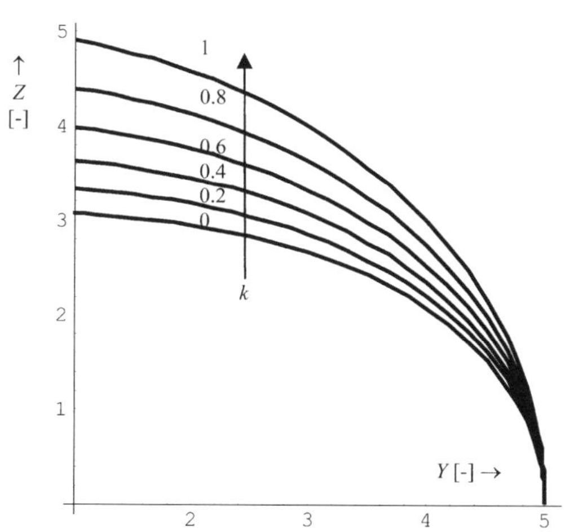


**Fig. 3.5** Optimal meridian profiles for various material orthotropy degrees $k$

### *Strength Analysis*

The individual layer stresses, as calculated in Eq 3.11, can serve as input parameters for a particular strength criterion that will lead to a minimally required laminate thickness (Ref 3.3, 3.4, 3.8, 3.9). With the expression for the meridian profile function (Eq 3.18), the shell stress vector (underlined because it refers to the laminate as a whole) becomes:

$$\underline{\sigma}^{(d)} = \begin{Bmatrix} \sigma_m^{(d)} \\ \sigma_p^{(d)} \\ 0 \end{Bmatrix} = \frac{Pc}{2t^{(d)}} \begin{Bmatrix} \left(\frac{(1+r)Y_{eq}^3}{Y^2}\right)\left(\frac{k+Y^2-1}{k+Y_{eq}^2-1}\right)^{\frac{k+1}{2}} \\ \left(\frac{(1+r)Y_{eq}^3}{Y^2}\right)\left(\frac{1+k\left(Y^2-1\right)}{\left(Y^2-1\right)+k}\right)\left(\frac{k+Y^2-1}{k+Y_{eq}^2-1}\right)^{\frac{k+1}{2}} \\ 0 \end{Bmatrix}$$

$$= \frac{Pc}{2t^{(d)}} \begin{Bmatrix} s_m^{(d)} \\ s_p^{(d)} \\ 0 \end{Bmatrix} = \frac{Pc}{2t^{(d)}} \underline{\mathbf{s}}^{(d)}$$

$$\underline{\boldsymbol{\varepsilon}}^{(d)} = \frac{Pc}{2t^{(d)}} \underline{\mathbf{C}}^{(d)} \cdot \underline{\mathbf{s}}^{(d)},$$

$$\underline{\mathbf{C}}^{(d)} = \left[\mathbf{S}^{(d)}\right]^{-1} = \left[\frac{1}{2}(\mathbf{S}(\alpha)+\mathbf{S}(-\alpha))\right]^{-1}$$

(Eq 3.20)

where $k_e$ is replaced by $k$, and the superscript (d) stands for the dome of the vessel. At $Y = Y_{eq}$, the stresses and strains of the shell become:

$$\underline{\sigma}_{eq}^{(d)} = \begin{Bmatrix} \sigma_m^{(d)} \\ \sigma_p^{(d)} \\ 0 \end{Bmatrix} = (1+r)Y_{eq}\left(\frac{Pc}{2t^{(d)}}\right) \begin{Bmatrix} 1 \\ \frac{1+k\left(Y_{eq}^2-1\right)}{\left(Y_{eq}^2-1\right)+k} \\ 0 \end{Bmatrix}$$

$$\underline{\boldsymbol{\varepsilon}}_{\text{eq}}^{(\text{d})} = \begin{Bmatrix} \varepsilon_{\text{m,eq}}^{(\text{d})} \\ \varepsilon_{\text{p,eq}}^{(\text{d})} \\ 0 \end{Bmatrix} = \underline{\mathbf{C}}^{(\text{d})} \cdot \underline{\boldsymbol{\sigma}}_{\text{eq}}^{(\text{d})} \qquad \text{(Eq 3.21)}$$

At the lower end of the dome (equator), the meridional curvature $k_{\text{m}}$ is not yet zero; hence, $R_{\text{m}}$ (= $1/k_{\text{m}}$) is not infinite. Otherwise, from Eq 3.5 one could immediately deduce that $\sigma_{\text{p}}^{(\text{d})} = 2\sigma_{\text{m}}^{(\text{d})}$. However, this applies only on a pure cylindrical shell. The parallel stress at the very end of the dome is equal to that of the cylindrical parallel stress if and only if:

$$\sigma_{\text{p}}^{(\text{c})} = 2\sigma_{\text{m}}^{(\text{d})} \rightarrow \frac{1 + k\left(Y_{\text{cyl}}^2 - 1\right)}{\left(Y_{\text{cyl}}^2 - 1\right) + k} = 2 \rightarrow Y_{\text{cyl}} = \sqrt{3}\sqrt{\frac{1-k}{2-k}} \qquad \text{(Eq 3.22)}$$

where the superscript (c) stands for cylinder. For $k = 0$, a value of $Y_{\text{cyl}} = (3/2)^{1/2}$ is obtained, and therefore $\alpha = 53.47°$. The layer stresses (not underlined) at the dome are given by:

$$\boldsymbol{\sigma}^{(\text{d})} = \begin{Bmatrix} \sigma_1^{(\text{d})} \\ \sigma_2^{(\text{d})} \\ \tau_{12}^{(\text{d})} \end{Bmatrix} = \frac{Pc}{2t^{(\text{d})}}\left[\mathbf{S}_0 \cdot \left(\mathbf{M}^{\text{T}}\right)^{-1} \cdot \underline{\mathbf{C}}^{(\text{d})}\right] \cdot \underline{\mathbf{s}}^{(\text{d})} = \frac{Pc}{2t^{(\text{d})}} \mathbf{T}^{(\text{d})} \cdot \underline{\mathbf{s}}^{(\text{d})} \qquad \text{(Eq 3.23)}$$

The layer orientations $\{+\alpha, -\alpha\}$ are a function of $Y$ (Eq 3.13). For the determination of the minimal required thickness, one can, for example, employ the Tsai-Hill criterion (Ref 3.7). This results in:

$$\left(\frac{Pc}{2t^{(\text{d})}}\right)^2 \left[\left(\frac{t_{11}^{(\text{d})} s_{\text{m}}^{(\text{d})} + t_{12}^{(\text{d})} s_{\text{p}}^{(\text{d})}}{S_1}\right)^2 + \left(\frac{t_{21}^{(\text{d})} s_{\text{m}}^{(\text{d})} + t_{22}^{(\text{d})} s_{\text{p}}^{(\text{d})}}{S_2}\right)^2 + \right.$$
$$- \frac{\left(t_{11}^{(\text{d})} s_{\text{m}}^{(\text{d})} + t_{12}^{(\text{d})} s_{\text{p}}^{(\text{d})}\right)\left(t_{21}^{(\text{d})} s_{\text{m}}^{(\text{d})} + t_{22}^{(\text{d})} s_{\text{p}}^{(\text{d})}\right)}{S_1 S_2}$$
$$\left. + \left(\frac{t_{31}^{(\text{d})} s_{\text{m}}^{(\text{d})} + t_{32}^{(\text{d})} s_{\text{p}}^{(\text{d})}}{S_3}\right)^2\right] \le 1$$

$$\rightarrow \left(\frac{Pc}{2t^{(\text{d})}}\right)^2 \underline{H}^{(\text{d})} \le 1 \rightarrow t^{(\text{d})} = \frac{Pc}{2}\sqrt{\underline{H}^{(\text{d})}} \qquad \text{(Eq 3.24)}$$

where $t_{ij}$ are elements of the 3 × 3 $\mathbf{T}^{(\text{d})}$ matrix (Eq 3.23). The terms $\{S_1, S_2\}$ are the tensional strengths in the fiber direction and vertical to it, respectively. The letter $S_3$ represents the in-plane shear strength. For every $Y\,[Y_{\text{min}}, Y_{\text{eq}}]$, the winding process should result in a dome thickness at least equal to $t^{(\text{d})}$.

A similar analysis can be set up for the cylindrical part (Fig. 3.4). Because this part is additionally covered by hoop fibers (for reasons of strain compatibility), the stiffness matrix becomes:

$$\left(\begin{array}{l} \underline{\mathbf{S}}^{(\text{c})} = \left(\dfrac{t_{\text{p}}^{(\text{c})}}{t_{\text{p}}^{(\text{c})} + t_{\text{h}}^{(\text{c})}}\right)\mathbf{S}_{\text{p}}^{(\text{c})} + \left(\dfrac{t_{\text{h}}^{(\text{c})}}{t_{\text{p}}^{(\text{c})} + t_{\text{h}}^{(\text{c})}}\right)\mathbf{S}_{\text{h}}^{(\text{c})} \\ t_{\text{h}}^{(\text{c})} = ft_{\text{p}}^{(\text{c})} = ft^{(\text{d})}(\text{at } Y = Y_{\text{eq}}) = ft_{\text{eq}}^{(\text{d})} \\ \alpha_{\text{p}}^{(\text{c})} = \alpha(\text{at } Y = Y_{\text{eq}}) = \alpha_{\text{eq}} \\ \alpha_{\text{h}}^{(\text{c})} = \pi/2 \end{array}\right) \rightarrow$$

$$\underline{\mathbf{S}}^{(\text{c})} = \left(\frac{1}{1+f}\right)\left[\frac{1}{2}\left(\mathbf{S}(\alpha_{\text{eq}}) + \mathbf{S}(-\alpha_{\text{eq}})\right)\right] + \left(\frac{f}{1+f}\right)\mathbf{S}(\pi/2)$$
$$2f = \text{positive integer} \qquad \text{(Eq 3.25)}$$

where "p" stands for the polar windings (in essence, a continuation of the dome windings), and "h" stands for the hoop windings, with $\alpha$ = 90°. Because a single closed polar layer is an angleply, a single hoop layer will have ½ of the polar thickness and requires that $2f$ (and not $f$) should be an integer. For the cylindrical part, the shell stresses are:

$$\underline{\boldsymbol{\sigma}}^{(\text{c})} = \begin{Bmatrix} \sigma_{\text{m}}^{(\text{c})} \\ \sigma_{\text{p}}^{(\text{c})} \\ 0 \end{Bmatrix} = \frac{Pc}{2t_{\text{eq}}^{(\text{d})}(1+f)} \begin{Bmatrix} (1+r)Y_{\text{eq}} \\ 2Y_{\text{eq}} \\ 0 \end{Bmatrix} = \frac{Pc}{2t_{\text{eq}}^{(\text{d})}(1+f)} \begin{Bmatrix} s_{\text{m}}^{(\text{c})} \\ s_{\text{p}}^{(\text{c})} \\ 0 \end{Bmatrix} = \frac{Pc}{2t_{\text{eq}}^{(\text{d})}(1+f)} \underline{\mathbf{s}}^{(\text{c})}$$

$$\underline{\boldsymbol{\varepsilon}}^{(c)} = \begin{Bmatrix} \varepsilon_m^{(c)} \\ \varepsilon_p^{(c)} \\ 0 \end{Bmatrix} = \underline{\mathbf{C}}^{(c)} \cdot \underline{\boldsymbol{\sigma}}^{(c)}, \qquad \underline{\mathbf{S}}^{(c)} = \left[\underline{\mathbf{C}}^{(c)}\right]^{-1}$$

(Eq 3.26)

Because $t_{eq}^{(d)}$ is given by Eq 3.24, a determination of the minimum value for $f$ that will ensure adequate strength for both the polar ($f_p$) and hoop layers ($f_h$) must be made. The stresses in an individual polar layer are:

$$\boldsymbol{\sigma}_p^{(c)} = \frac{Pc}{2t_{eq}^{(d)}(1+f)}\left[\mathbf{S}_0 \cdot \left(\mathbf{M}_p^{T}\right)^{-1} \cdot \underline{\mathbf{C}}^{(c)}\right] \cdot \underline{\mathbf{s}}^{(c)} = \frac{Pc}{2t_{eq}^{(d)}(1+f)}\mathbf{T}_p^{(c)} \cdot \underline{\mathbf{s}}^{(c)} \qquad \text{(Eq 3.27)}$$

with $\mathbf{M}_p = \mathbf{M}(\alpha_{eq})$. As the number of hoop layers is increased, the stress level in the polar layers will be decreased. To ensure that this stress level is below the strength of these polar layers, there is a minimum number of hoop layers, $f_p$, needed. This number can be determined as follows:

$$\left(\frac{Pc}{2t_{eq}^{(d)}\left(1+f_p\right)}\right)^2\left[\left(\frac{t_{p,11}^{(c)}s_m^{(c)}+t_{p,12}^{(c)}s_p^{(c)}}{S_1}\right)^2 + \left(\frac{t_{p,21}^{(c)}s_m^{(c)}+t_{p,22}^{(c)}s_p^{(c)}}{S_2}\right)^2 + -\frac{\left(t_{p,11}^{(c)}s_m^{(c)}+t_{p,12}^{(c)}s_p^{(c)}\right)\left(t_{p,21}^{(c)}s_m^{(c)}+t_{p,22}^{(c)}s_p^{(c)}\right)}{S_1S_2} + \left(\frac{t_{p,31}^{(c)}s_m^{(c)}+t_{p,32}^{(c)}s_p^{(c)}}{S_3}\right)^2\right] \le 1$$

$$\rightarrow \left(\frac{Pc}{2t_{eq}^{(d)}\left(1+f_p\right)}\right)^2 H_p^{(c)} \le 1$$

$$\rightarrow f_p \ge \frac{Pc}{2t_{eq}^{(d)}}\sqrt{H_p^{(c)}} - 1 \qquad \text{(Eq 3.28)}$$

where $t_{p,ij}^{(c)}$ are elements of the $\mathbf{T}_p^{(c)}$ matrix. The obtained $f_p$ number is then rounded to the next multiple of a single layer (in practice: 0.2 to 0.5 mm). For example, by assuming layer thickness of 0.3 mm, a hoop laminate thickness of 2.31 would become 2.4 mm. By just replacing the "p" index in the $t_{ij}$ elements with "h," the minimum number of hoop layers ($f_h$) providing sufficient hoop strength is calculated.

Finally, the combined minimum number of hoop layers becomes:

$$f_{min} = \max\left[f_p, f_h\right] \qquad \text{(Eq 3.29)}$$

Next, to avoid bending effects at the cylinder-dome intersection, the associated circumferential strains of both dome and cylinder must be equal at exactly that location, $\varepsilon_{p,eq}^{(d)} = \varepsilon_p^{(c)}$ (Ref 3.10). From Eq 3.21 and 3.26, Eq 3.30 is developed:

$$\left(\frac{Pc}{2t_{eq}^{(d)}}\right)(1+r)Y_{eq}\left[\underline{C}_{21}^{(d)} + \underline{C}_{22}^{(d)}\left(\frac{1+k_{match}\left(Y_{eq}^2-1\right)}{\left(Y_{eq}^2-1\right)+k_{match}}\right)\right] = \left(\frac{Pc}{2t_{eq}^{(d)}}\right)\left(\frac{1}{1+f_{min}}\right)\left[(1+r)Y_{eq}\underline{C}_{21}^{(c)} + 2Y_{eq}\underline{C}_{22}^{(c)}\right]$$

$$\rightarrow \underline{C}_{21}^{(d)} + \underline{C}_{22}^{(d)}\left(\frac{1+k_{match}\left(Y_{eq}^2-1\right)}{\left(Y_{eq}^2-1\right)+k_{match}}\right) = \underline{C}_{21}^{(c)}\left(\frac{1}{1+f_{min}}\right) + \underline{C}_{22}^{(c)}\left(\frac{2}{(1+r)\left(1+f_{min}\right)}\right) \rightarrow$$

$$\rightarrow k_{match} = \frac{\left(1+f_{min}\right)(1+r)\left[\underline{C}_{22}^{(d)} + \underline{C}_{21}^{(d)}\left(Y_{eq}^2-1\right)\right] - \left[\underline{C}_{21}^{(c)}(1+r) + 2\underline{C}_{22}^{(c)}\right]\left(Y_{eq}^2-1\right)}{\left[\underline{C}_{21}^{(c)}(1+r) + 2\underline{C}_{22}^{(c)}\right] - \left(1+f_{min}\right)(1+r)\left[\underline{C}_{22}^{(d)} + \underline{C}_{21}^{(d)}\left(Y_{eq}^2-1\right)\right]}$$

(Eq 3.30)

which is a function of $\{Y_{eq}, r, f_{min}\}$. With the new $k$ parameter, $k_{match}$, the strength must be rechecked by first calculating the new $t_{eq}^{(d)}$ and then using it for the determination of $f_{min}$. Should this new $f_{min}$ be the same as the previous one, then the design is final; otherwise, this iteration loop must be repeated. With this last step, the meridian profile becomes:

$$Z(Y) = \int_{\tau=Y_{min}}^{Y} Z'(\tau)d\tau$$

where $Z'(\tau) =$

$$\frac{\tau\left(\tau^2 + rY_{eq}^2\right)}{\sqrt{\left(\dfrac{k_{match}+\tau^2-1}{k_e+Y_{eq}^2-1}\right)^{k_{match}+1}(1+r)^2Y_{eq}^6 - \tau^2\left(\tau^2+rY_{eq}^2\right)^2}}$$

(Eq 3.31)

A pressure vessel based on this equation satisfies all strength and deformation compatibility constraints at approximately minimum weight.

## Winding Patterns

### *A Game of Integers*

The structural design procedure for pressure vessels dictates a certain minimum thickness distribution over the extended meridian profile (dome + cylinder). This thickness must be built up from individual windings running over the vessel surface (Ref 3.11). At the equatorial periphery, $2\pi c Y_{eq}$, a number, $n$, of roving widths will fit in a single layer ($n$ is not necessarily an integer number). As the rovings cross the equatorial periphery under an angle $\alpha_{eq}$ (Fig. 3.1, 3.4), the effective width is:

$$B_{eff} = \frac{B}{\cos(\alpha_{eq})} = \frac{BY_{eq}}{\sqrt{Y_{eq}^2 - 1}}, \quad B = \frac{b}{c} \qquad \text{(Eq 3.32)}$$

where $B$ is the dimensionless roving width (actual width, $b$, divided by the polar opening radius, $c$). The roving thickness is $t = Tc$. The dimensionless cross-sectional area of a single roving is $BT$ for dry winding. For wet winding, this must be multiplied by a parameter $w > 1$ that accounts for the resin inclusion. As a first attempt, this parameter can be approximated by $1/v_f$, where $v_f$ is the fiber volume fraction. In practice, however, the parameter $w$ will be less than $1/v_f$ due to the effects of compaction and consolidation. Every single complete winding will result in two rovings crossing the equatorial periphery (Fig. 3.1, 3.4). Taking this into account, the required number of windings ($nd$) is:

$$2ndwB_{eff}T = 2\pi Y_{eq}T_{eq} \rightarrow nd = \mathrm{CE}\left[\left(\frac{\pi}{w}\right)\left(\frac{\sqrt{Y_{eq}^2 - 1}}{B}\right)\right] \qquad \text{(Eq 3.33)}$$

where CE stands for ceiling, and $t_{eq}^{(d)} = cT_{eq}$. The function CE gives the smallest integer greater than or equal to the argument. For example, the number 2091.3 would become 2092. This minor overprediction is needed because $T_{eq}$ represents the minimum wall thickness being just capable of providing adequate strength. Note that while $d$ is an integer, $n$ does not need to be, for example, $d \times n = 4 \times 127.5 = 255$.

The construction of a leading winding pattern is illustrated in Fig. 3.6, where a schematic topview of the vessel is presented, earlier

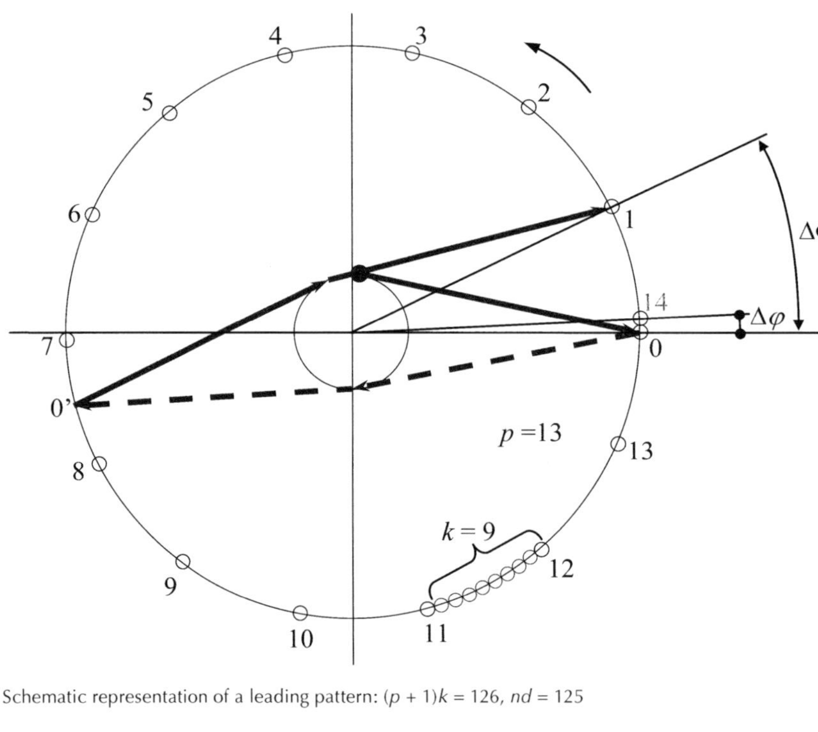


**Fig. 3.6** Schematic representation of a leading pattern: $(p + 1)k = 126$, $nd = 125$

depicted three dimensionally in Fig. 3.1. The winding process starts at the gray spot on the periphery of the polar opening and proceeds to position 0, where it crosses the equatorial periphery. Next, by running over the bottom side of the vessel (dotted line), it crosses the equator again at 0′, but this time with the opposite winding angle. As the winding evolves, it returns to the right side, position 1, and so on. The angular difference between positions 0 and 1 is $\Delta\Phi$. The angle occupied by a single roving width, $B_{\text{eff}}$, is indicated as $\varphi$. The diameters of the small circles around the periphery are equal to $B_{\text{eff}}$. The process will continue in counterclockwise direction. At winding 14, the roving is placed exactly ahead of winding 0. The number ($p$) of $\Delta\Phi$ jumps fitting in the equatorial periphery is 13. The number of $\Delta\varphi$ intervals ($k$) fitting in $\Delta\Phi$ is 9. The number of $\Delta\varphi$ elements ($nd$) that fit in the equatorial periphery is 125.

It is clear that $14 \times 9 - 125 = 1$. This situation corresponds to one closed layer. For two layers, the rovings should overlap for 50% (to create some staggering); hence, $14 \times 9 - 125 = 1/2$, which is, at first sight, impossible. The correct formulation is $14 \times 9 \times 2 - 251 = 1$, implying that the number of rovings that have to fit in the periphery is not 125 but 125.5 (for a single layer). This is why the total number of rovings is given by $nd$, where $d$ stands for the number of closed layers. In conclusion, for $d > 1$, $n$ does not have to be an integer, but it always is a multiple of $1/d$:

$$(p+1)kd - nd = 1 \qquad \text{(Eq 3.34)}$$

For a lagging pattern (Fig. 3.7), the 14th winding is located exactly before winding 0. This time, the number of $\Delta\varphi$ elements fitting in the equatorial periphery is 14 (and not 13, as in the case of the leading pattern). A successful single-layer pattern is now established when $14 \times 9 - 127 = -1$. In this case, a roving needs to be selected that is $^{125}/_{127}$% narrower than $B_{\text{eff}}$ for leading winding. In general, for multiple layers, a formula for a lagging pattern can thus be written as:

$$pkd - nd = -1 \qquad \text{(Eq 3.35)}$$

The realization of a pattern is entirely dependent on the angular difference between two neighboring windings and the roving dimensions $\{B_{\text{eq}}, T\}$. The angle $\Delta\Phi$ is measured on the $Y_{\text{eq}}$ periphery between two adjacent wind-

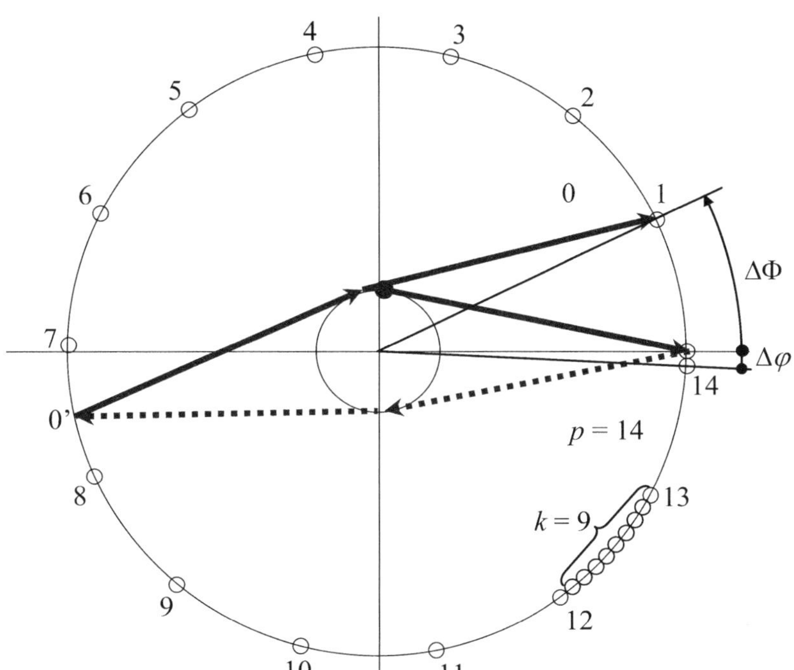


**Fig. 3.7** Schematic representation of a lagging pattern: $pk = 126$, $nd = 127$

ings (Fig. 3.6, 3.7). The unbounded value of this angle is given by:

$$\Phi_{\text{tot}} = 2\Phi^{(c)} + 4\Phi^{(d)}$$

$$\Phi^{(c)} = \frac{H}{Y_{\text{eq}}}\tan\alpha_{\text{eq}}$$

$$\Phi^{(d)} = \int_{Y_{\min}}^{Y_{\text{eq}}} \frac{\sqrt{1+Z'^2(Y)}}{Y}\tan(\alpha(Y))\,dY \qquad \text{(Eq 3.36)}$$

where $\Phi^{(c)}$ is the $\phi$ propagation of the roving over the cylindrical part, for example, from top to bottom; $\Phi^{(d)}$ is the $\phi$ propagation over the dome, in one direction and on one side. The total propagation can be either positive or negative and must be limited within $2\pi$ (Ref 3.6):

$$\Delta\Phi_{\text{tot}} = \min\left\{\left|\text{mod}_{(2\pi)}(\Phi_{\text{tot}})\right|, \left|\text{mod}_{(-2\pi)}(\Phi_{\text{tot}})\right|\right\} \qquad \text{(Eq 3.37)}$$

Next, the effective roving width, $B_{\text{eff}}$ (Eq 3.32), is now converted into radians:

$$\Delta\varphi = \frac{B_{\text{eff}}}{Y_{\text{eq}}} = \frac{B}{\sqrt{Y_{\text{eq}}^2 - 1}} \qquad \text{(Eq 3.38)}$$

Plugging this expression in, Eq 3.33 immediately gives:

$$d = \text{CE}\left[\frac{T_{\text{eq}}}{2wT}\right] \qquad \text{(Eq 3.39)}$$

which is an obvious result. From the previous discussion, the pattern parameters $p$ and $k$ can be defined as (Ref 3.6):

$$p = \text{IP}\left[\frac{2\pi}{\Delta\Phi}\right]$$

$$k = \text{IP}\left[\frac{\Delta\Phi}{\Delta\varphi}\right] \qquad \text{(Eq 3.40)}$$

where IP stands for integer part; this function gives the biggest integer contained in the argument, for example, IP[7.8] = 7.

### *Pattern Search*

A preliminary investigation for possible patterns can be carried out as follows. Translating Eq 3.34 and 3.35 in goniometric quantities leads to:

$$(p+1)\Delta\Phi - 2\pi = \frac{\Delta\varphi}{d} \qquad \text{leading}$$

$$p\Delta\Phi - 2\pi = -\frac{\Delta\varphi}{d} \qquad \text{lagging} \qquad \text{(Eq 3.41)}$$

Substitution of Eq 3.38 and 3.39 into Eq 3.33 (temporarily assuming values that exactly produce integer quantities) provides an alternative formulation for $\Delta\varphi$:

$$nd = \left(\frac{\pi}{w}\right)\left(\frac{\sqrt{Y_{\text{eq}}^2 - 1}}{B}\right)\left(\frac{T_{\text{eq}}}{T}\right) = \frac{2\pi d}{\Delta\varphi} \rightarrow \Delta\varphi = \frac{2\pi}{n} \qquad \text{(Eq 3.42)}$$

Plugging this result into Eq 3.41 leads to:

$$\Delta\Phi_{\text{leading}} = \frac{2\pi}{p+1}\left(1 + \frac{1}{nd}\right)$$

$$\Delta\Phi_{\text{lagging}} = \frac{2\pi}{p}\left(1 - \frac{1}{nd}\right) \qquad \text{(Eq 3.43)}$$

Comparison of these values with Eq 3.37 will generate a preliminary judgment about how close the vessel-determined $\Delta\Phi_{\text{tot}}$ is to $\Delta\Phi_{\text{leading}}$ or $\Delta\Phi_{\text{lagging}}$. In practice, the allowable difference depends on the relative size of the vessel and the available friction. After all, every correction in the spindle rotation of the winding machine will induce a deviation from the geodesic path. As a rule of thumb, the angular difference should not be more than 0.5°.

The next step is to determine a range of feasible (or available) $B$ and $T$ values. For every $\{B, T\}$ combination, the parameters $\{p, d, k, nd\}$ are calculated and checked whether they satisfy conditions in Eq 3.34 or 3.35. When the designer is not limited to particular $B$ and $T$ values, the best option is to discretize a given interval $[B_{\min}, B_{\max}]$ and a realistic range for $[T_{\min}, T_{\max}]$ and to evaluate every combination on pattern existence. In this way, the (classical) method of continued fractions for solving Eq 3.34 and 3.35 is not needed (Ref 3.11).

## Laminate Thickness Distribution

### *Classic Method*

Up to now, the winding pattern has been designed for providing at least the required thickness at the equator of the vessel. Assuming that such a pattern has been selected, the resulting equatorial thickness can be written as (see also Eq 3.33):

$$nd = \left(\frac{\pi}{w}\right)\left(\frac{\sqrt{Y_{eq}^2 - 1}}{B}\right)\left(\frac{T_{eq}}{T}\right) \rightarrow T_{eq}$$

$$= \left(\frac{nd}{\pi}\right)\frac{wBT}{\sqrt{Y_{eq}^2 - 1}} \qquad \text{(Eq 3.44)}$$

Proceeding from the equator of the pressure vessel to the polar areas, it is expected that the laminate thickness will grow rapidly; the same number of rovings must be distributed over a smaller periphery ($Y < Y_{eq}$) while they tend to lay in a more parallel fashion, resulting in even more overlap (the winding angle, $\alpha$, increases). This mechanism is captured by a simple consideration: a number of rovings are distributed around a periphery with radius $Y$:

$$2ndw\left[\frac{B}{\cos(\alpha(Y))}\right]T = 2\pi Y T_{sm} \qquad \text{(Eq 3.45)}$$

where $T_{sm}$ is the smeared out thickness over a periphery on "altitude" $Y$. Substitution of Eq 3.45 into this expression gives the classical expression (Ref 3.5):

$$T_{sm}^{cl} = T_{eq}\sqrt{\frac{Y_{eq}^2 - 1}{Y^2 - 1}} \qquad \text{(Eq 3.46)}$$

As pointed out in Ref 3.5 and 3.8, this expression leads to infinite thickness at the polar periphery (where $Y = 1$). This occurs because it is assumed that the winding angle is constant over the entire roving width and that the roving overlap takes place over the same radius $Y$.

### *Exact Solution*

Without any derivation, the exact solution that considers these effects is provided, as published in Ref 3.8:

$$T_{sm}^{exact} = T_{eq}\left[\left(\frac{Y_{eq}\sqrt{1+[Z'(Y)]^2}}{B}\right)\arccos\left(\frac{1}{Y}\right)\right]\cos(\alpha_{eq}), \quad 1 \le Y \le 1+B$$

$$T_{sm}^{exact} = T_{eq}\left\{\arcsin\left[\frac{1}{Y}\left(1+\frac{B}{\sqrt{1+[Z'(Y)]^2}}\right)\right] - \arcsin\left(\frac{1}{Y}\right)\right\}\cos(\alpha_{eq}), \quad 1+B \le Y \le Y_{eq}$$

(Eq 3.47)

Expansion of this solution in a Taylor series leads to an approximation where the first-order term is equal to Eq 3.46 (Ref 3.8). As depicted in Fig. 3.8, the exact solution shows a sharp peak at $Y = 1 + B$; although correct from a mathematical

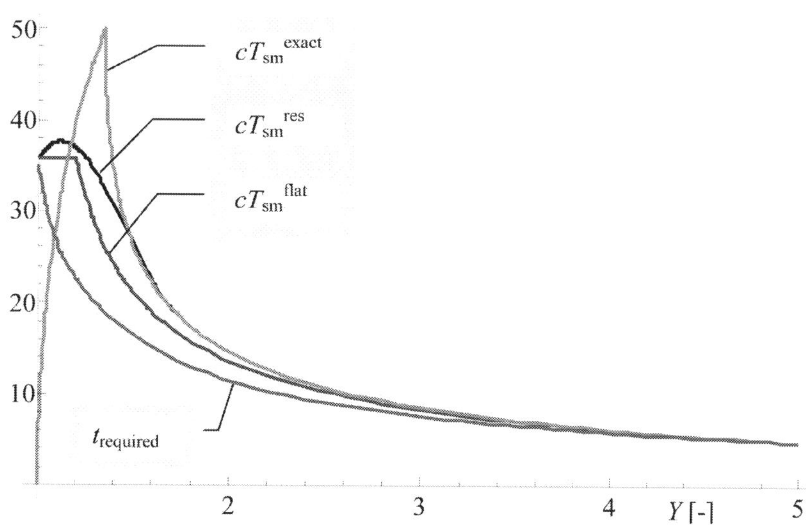


**Fig. 3.8** Comparison of various methods for the laminate thickness distribution in millimeters. $Y_{eq} = 5$, $k = 0.18$, $r = 0$, $c = 40$ mm (1.6 in.)

point of view, this phenomenon does not occur in practice due to compaction and consolidation effects, fiber slipping, and air entrapment (particularly in cases where the fibers, placed on the equatorial periphery, do not lie exactly next to each other but show some overlap). Therefore, to approximate the laminate thickness, a "smoothing" of this result is preferred.

### *Flat Solution*

As an intermediate step for obtaining the smooth thickness distribution, some attention must first be paid to an alternative way for approximating the thickness close to the polar area (Ref 3.4). This is done by adopting the thickness distribution as given in Eq 3.46, with a small modification within the ring area determined by $[Y_w, 1]$. The surface area, occupied by this roving, is given by Fig. 3.9:

$$S_{polar} = \omega(c+b)^2 - c^2 \tan\omega = c^2\left[(1+B)^2 \arccos\left(\frac{1}{1+B}\right) - \sqrt{B(2+B)}\right] \quad \text{(Eq 3.48)}$$

Smearing out the thickness generated by $2ndw$ rovings over the ring area, Eq 3.49 is obtained:

$$2ndwTS_{polar} = T_{polar}\left[\pi(c+b)^2 - \pi c^2\right] = T_{polar}\pi b(2c+b) \quad \text{(Eq 3.49)}$$

Substitution of Eq 3.44 into this equality finally leads to:

$$T_{polar} = T_{eq}\frac{\sqrt{Y_{eq}^2-1}\left[(1+B)^2 \arccos\left(\frac{1}{1+B}\right) - \sqrt{B(2+B)}\right]}{B^2(2+B)} \quad \text{(Eq 3.50)}$$

The thickness at the polar area drops exponentially with increasing $B$. The only parameters required to determine $T_{polar}$ are the vessel aspect ratio, $Y_{eq}$, and the normalized roving bandwidth, $B$ (= $b/c$).

The $Y$ value for which the "classical" thickness distribution (Eq 3.46) becomes equal to $T_{polar}$ can be determined by setting Eq 3.46 equal to Eq 3.50 and solving for $Y$. The solution, $Y_w$, is:

$$Y_w = \sqrt{1 + \frac{B^4(2+B)^2}{\left[(1+B)^2 \arccos\left(\frac{1}{1+B}\right) - \sqrt{B(2+B)}\right]^2}} \quad \text{(Eq 3.51)}$$

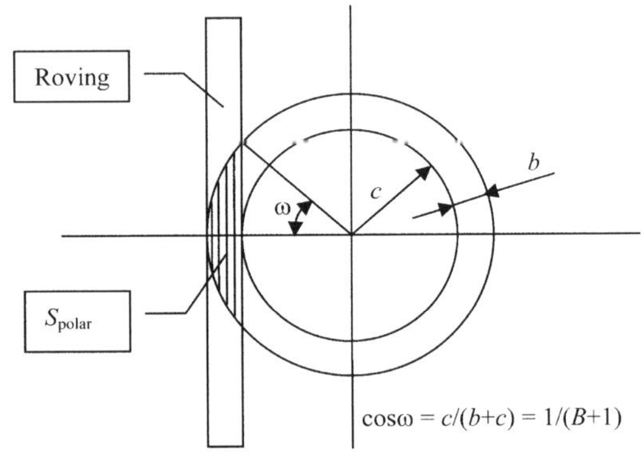


**Fig. 3.9** Roving placement, adjacent to the polar periphery of a pressure vessel (top view)

With this result, the total distribution over the $[1, Y_{eq}]$ domain becomes:

$$T_{sm}^{flat} = T_{polar}, \quad 1 \le Y \le Y_w$$

$$T_{sm}^{flat} = T_{eq}\sqrt{\frac{Y_{eq}^2-1}{Y^2-1}}, \quad Y_w \le Y \le Y_{eq} \quad \text{(Eq 3.52)}$$

It seems that this thickness schematization is able to approximate reality in a very satisfactory way (Fig. 3.8). In particular, the value it provides for $Y = 1$ is very close to measured thicknesses (Ref 3.5, 3.8). An important observation is that the $T_{sm}^{flat}$ distribution always provides lower values than $T_{sm}^{exact}$, which is the exact solution. For $Y \in [1 + B, 1 + 2B]$, it is found that $T_{sm}^{exact}$ is closer to reality. However, for $Y \in [1, 1 + B]$, the $T_{sm}^{flat}$ approximation is preferred.

### *Polynomial Approximation*

A method combining all advantages of $T_{sm}^{cl}$, $T_{sm}^{exact}$, and $T_{sm}^{flat}$ is a polynomial approximation of a combination of them (Ref 3.4, 3.8), hence defining:

$$T_{sm}^{p} = a_0 + a_1Y + a_2Y^2 + a_3Y^3 \quad \text{(Eq 3.53)}$$

According to Ref 3.8, it is justified to assume the validity range of the polynomial as $Y \in [1, 1 + 2B]$.

To determine the unknown coefficients, the following continuity and equivalence conditions are used:

- $Y = 1$: $T_{sm}^{p} = T_{polar}$
- $Y = 1 + 2B$: $T_{sm}^{p} = T_{sm}^{exact}$
- $Y = 1 + 2B$: $(T_{sm}^{p})' = T_{sm}^{exact}$ (′ means differentiation w.r.t. $Y$)

- $$\int_{Y=1}^{Y=1+2B} T_{sm}^{p}\,dY = \int_{Y=1}^{Y=1+2B} T_{sm}^{exact}\,dY$$

These conditions lead to the following system of equations:

$$\begin{Bmatrix} T_{polar} \\ T_{sm}^{exact}(1+2B) \\ \dfrac{d}{dY}\left(T_{sm}^{exact}\right)(1+2B) \\ \displaystyle\int_{Y=1}^{Y=1+2B} T_{sm}^{exact}\,dY \end{Bmatrix} = \mathbf{A}\cdot\begin{Bmatrix} a_0 \\ a_1 \\ a_2 \\ a_3 \end{Bmatrix} =$$

$$= \begin{bmatrix} 1 & 1 & 1 & 1 \\ 1 & 1+2B & (1+2B)^2 & (1+2B)^3 \\ 0 & 1 & 2(1+2B) & 3(1+2B)^2 \\ 2B & 2B(1+B) & \frac{2}{3}B\left[3+B(6+4B)\right] & \frac{2}{3}B\left[3+9B+6B^2(2+B)\right] \end{bmatrix}\cdot\begin{Bmatrix} a_0 \\ a_1 \\ a_2 \\ a_3 \end{Bmatrix} \quad \text{(Eq 3.54)}$$

The determinant is equal to $16B^6/3 \neq 0$, so the system always has a solution for nonzero roving width values. For convenience, the inverse of the matrix in Eq 3.54 is provided:

$$\mathbf{A}^{-1} = \begin{bmatrix} \dfrac{2B(1+2B)^2}{4B^3} & \dfrac{4+3B(5+4B)}{4B^3} & -\dfrac{(1+B)(1+2B)}{2B^2} & -\dfrac{3(1+2B)^2}{4B^4} \\ -\dfrac{3(1+B)(1+2B)}{2B^3} & -\dfrac{3(2+B)(1+2B)}{2B^3} & 1+\dfrac{3+6B}{2B^2} & \dfrac{3[3+4B(2+B)]}{4B^4} \\ \dfrac{6+9B}{4B^3} & \dfrac{3(4+5B)}{4B^3} & -\dfrac{3(1+B)}{2B^2} & -\dfrac{3(3+4B)}{4B^4} \\ -\dfrac{1}{2B^3} & -\dfrac{1}{B^3} & \dfrac{1}{2B^2} & \dfrac{3}{4B^4} \end{bmatrix} \quad \text{(Eq 3.55)}$$

Once the coefficients are known, the resulting thickness distribution becomes:

$$\begin{aligned} T_{sm}^{res} &= T_{sm}^{p} && 1 \le Y \le 1+2B \\ T_{sm}^{res} &= T_{sm}^{exact} && 1+2B \le Y \le Y_{eq} \end{aligned} \quad \text{(Eq 3.56)}$$

In Fig. 3.8, a graph of the thickness distribution according to Eq 3.56 is provided (line labeled $cT_{sm}{}^{res}$). Observe that this distribution lies between $T_{sm}{}^{exact}$ and $T_{sm}{}^{flat}$. According to Vasiliev et al. (Ref 3.8) and Koussios et al. (Ref 3.5), this is probably the most realistic representation for the laminate thickness as it results from filament winding.

With the availability of these thickness approximations, it is now possible to check if the shell thickness, as resulting from the winding process, is at least equal to the thickness distribution demanded by Eq 3.24 for every $Y$ belonging to the interval [1, $Y$]. Obviously, at the equator, the "structural" thickness is the same or a bit less than the "winding" thickness; this has formed a hard constraint for setting up the winding patterns. Without guaranteeing sufficient thickness over the entire $Y$ range, a point that should certainly be checked is $Y$ = 1. The most pessimistic value for the winding thickness at that point is $T_p$ (Eq 3.50). An increasing roving width $B$ speeds up the winding process (less windings needed) but reduces the structural performance due to increased deviations from the ideal, zero-width, geodesic path. Moreover, for sufficiently large $B$, the $T_p$ value may not be reached by the winding process. In this case, the upper bound for $B$ is obtained by solving the following equality:

$$Y = 1: \quad \frac{t^{(d)}}{c} = T_p(B) \quad \text{(Eq 3.57)}$$

For $k$ = 0 (netting approach), the fibers are equally loaded in tension everywhere; therefore, the winding thickness will always be sufficient. Note, however, that in this case the minimum radius, $Y_{min}$, is bigger than the $Y_{min}$ provided by $k > 0$, so the nonoptimal area around the pole is also bigger. In practice, however, the fibers do have to approach the polar opening until the winding angle becomes 90° at $Y$ = 1. In this case, the upper bound for $B$ counts for the (larger) discrepancy between $Y_{min}$ and 1 and for path deviations of the real roving as compared to the theoretical one of zero width.

The lower boundary for $B$ is determined by production issues: the lower the $B$ value, the more windings needed. This leads to increased production time and, in general, increased thickness buildup at the polar areas, which can reduce the structural performance as well. However, these bounds usually have only a theoretical character because the stress state around the pole is dominated by the flange.

## Design Example

### *Parameters*

A pressure vessel is made by a specific resin-fiber combination ($v_f = 0.5$), with:

- $E_x = 8702$ ksi (60,000 MPa)
- $E_y = 1421$ ksi (9800 MPa)
- $v_{xy} = 0.3$
- $G_{xy} = 493$ ksi (3400 MPa)
- $S_{1,\text{ tension}} = 218$ ksi (1500 MPa)
- $S_{1,\text{ compression}} = -196$ ksi (–1350 MPa)
- $S_{2,\text{ tension}} = 6$ ksi (40 MPa)
- $S_{2,\text{ compression}} = -30$ ksi (–210 MPa)
- $T_{12} = 7$ ksi (50 MPa)

The design parameters are:

- $P = 1.5$ ksi (10 MPa)
- $c = 0.79$ in. (20 mm)
- $r = 0$
- $Y_{eq} = 10$
- $H = 2$

### *Structural Analysis*

The minimum allowable equatorial thickness is $t_{eq} = 0.19$ in. (4.8 mm); this requirement is based on $k = k_e = 0.202415$. To ensure sufficient strength for the cylindrical part, the minimum number of hoop layers, $f_{min}$, is equal to 2. With this number, the $k$ value that ensures circumferential strains compatibility for the dome and cylinder is $k_{match} = 0.188949$. With this new value, the number $f_{min}$ is recalculated. The new $f_{min}$ value is also equal to 2, so the design can now be frozen.

### *Winding Patterns*

For the obtained set of design variables $\{P, c, k_{match}, r, Y_{eq}, f\}$, the angular winding propagation is $\Delta\Phi = 9.9662°$. The following ranges are assumed for the dimensionless roving dimensions $\{B, T\}$:

- $B = \{10/200, 11/200, 12/200, \ldots, 200/200\}$
- $T = \{10/2000, 15/2000, \ldots, 50/2000\}$

The $\{\Delta\Phi_{leading}, \Delta\Phi_{lagging}\}$ values are very close to $\Delta\Phi_{tot}$, within 0.25°. For the given $\{B, T\}$ range, there must be a collection of feasible patterns providing the required $T_{eq}$ thickness (or, in other terms, ensuring that the number of windings, $nd$, is big enough to build up this thickness). For every $\{B, T\}$ combination, it is necessary to verify that the conditions for leading and lagging patterns are satisfied. The results for leading winding are summarized in Table 3.1, where the greatest common divisor for the obtained integer values should be equal to 1; otherwise, the pattern will be repeated, and at least two rovings will then be placed exactly on top of each other. In Table 3.2, possible lagging patterns are listed.

### *Acceptable Roving Dimensions*

To reduce production costs, a small number of windings ($nd$) is preferred. However, this implies a broader roving and therefore more deviations from the optimal path, leading to reduced structural performance. As the roving width increases, the laminate thickness at the polar area will generally decrease. The required thickness in that region ($Y = 1$) is $t_{d(Y=1)} = 2.399$ in. (60.925 mm). This implies that $B$ must be less than 0.477 or $b < 0.38$ in. (9.54 mm). Consequently, only the winding processes with $nd = 2404$ are acceptable (first row in both tables). However, by slightly increasing the corresponding $T$ value ($T = 0.0125$ for $nd = 2404$) to at least $(0.485/0.477) \times 0.0175 = 0.0178$, a dry roving thickness of 0.014 in. (0.356 mm) ($t = cT$) will provide the ability to use the second pattern (in both tables), with $nd = 443$ windings. Because the angular difference for the lagging pattern (second row in Table 3.2) is smaller, the best winding pattern is:

$b = 0.38$ in. (9.7 mm) (in practice: 0.39 in., or 10 mm)

$t = 0.014$ in. (0.356 mm)

$p = 36$, $k = 3$, $d = 4$, $nd = 443$

For thickness check:

Layer thickness = 0.028 in. (0.712 mm)

($w \approx 1/v_f = 2$)

$d = \text{CE}[T_{eq}/(2wT)] = \text{CE}\,[3.93] = 4$

For pattern check:

$p = 36 \quad k = 3 \quad nd = 443$

Verification: $36 \times 3 \times 4 - 443 = 442 - 443 = -1$

$\Delta\Phi_{tot} - \Delta\Phi_{leading}$ is negligible (0.0112274°).

**Table 3.1 Feasible leading patterns**

| *B* | *T* | *p* | *k* | *d* | *nd* | $\Phi_{tot} - \Phi_{leading}$, degrees | GCD(a) |
|---|---|---|---|---|---|---|---|
| 0.125 | 0.0125 | 37 | 13 | 5 | 2404 | 0.232422 | 1 |
| 0.485 | 0.0175 | 37 | 3 | 4 | 443 | 0.214506 | 1 |
| 0.68 | 0.015 | 37 | 2 | 5 | 369 | 0.210102 | 1 |
| 0.68 | 0.025 | 37 | 2 | 3 | 221 | 0.192444 | 1 |
| 0.815 | 0.0125 | 37 | 2 | 5 | 69 | 0.210102 | 1 |

(a) GCD, greatest common divisor for the obtained integer values

**Table 3.2 Feasible lagging patterns**

| *B* | *T* | *p* | *k* | *d* | *nd* | $\Phi_{tot} - \Phi_{leading}$, degrees | GCD(a) |
|---|---|---|---|---|---|---|---|
| 0.125 | 0.0125 | 36 | 13 | 5 | 2404 | −0.029641 | 1 |
| 0.485 | 0.0175 | 36 | 3 | 4 | 443 | −0.0112274 | 1 |
| 0.68 | 0.015 | 36 | 2 | 5 | 369 | −0.00670048 | 1 |
| 0.68 | 0.025 | 36 | 2 | 3 | 221 | +0.0114481 | 1 |
| 0.815 | 0.0125 | 36 | 2 | 5 | 369 | −0.00670048 | 1 |

Notice that this (small) $\Phi$ difference will become equal to $0.0112274 \times 443 = 4.97374°$ after completing the entire winding process. This angular deviation corresponds to almost two times the roving width (0.684 in., or 17.3616 mm). Therefore, because the spindle rotation of the winding machine dictates the winding pattern, the associated rotation between two consequent windings should always exactly match the $\Phi_{leading,lagging}$ value. This is why the spindle of a winding machine must allow for very accurate computer numerical control, up to at least four digits behind the decimal point (in degrees).

## REFERENCES


3.1 E.H. Baker, L. Kovalevsky, and F.L. Rish, *Structural Analysis of Shells*, McGraw-Hill Book Company, New York, 1972

3.2 W. Flügge, *Stresses in Shells*, Springer Verlag, Berlin/Heidelberg/New York, 1966

3.3 Th. de Jong, "A Theory of Filament Wound Pressure Vessels," Internal report LR-379, Structures and Materials Laboratory, Delft University of Technology, Delft, 1983

3.4 S. Koussios, O.K. Bergsma, and A. Beukers, "The Role of Material Anisotropy in the Optimal Design of Cylindrical Composite Pressure Vessels," 22nd Annual Conference of the American Society for Composites, Sept 2007 (Seattle, WA)

3.5 S. Koussios and A. Beukers, "Influence of Laminate Thickness Approximation Methods on the Performance of Optimal Filamentary Pressure Vessels," 23rd Annual Conference of the American Society for Composites, Sept 9–11, 2008 (Memphis, TN)

3.6 S. Koussios, "Design of Cylindrical Composite Pressure Vessels: Integral Optimisation," 17th International Conference on Composite Materials, July 2009 (Edinburgh, U.K.)

3.7 I.M. Daniel and O. Ishai, *Engineering Mechanics of Composite Materials*, Oxford University Press, New York/Oxford, 2006

3.8 V.V. Vasiliev, A.A. Krikanov, and A.F. Razin, New Generation of Filament-Wound Composite Pressure Vessels for Commercial Applications, *Compos. Struct.*, Vol 62, 2003, p 449–459

3.9 C.-C. Liang, H.-W. Chen, and C.-H. Wang, Optimum Design of Dome Contour for Filament Winding Composite Pressure Vessels Based on a Shape Factor, *Compos. Struct.*, Vol 58, 2002, p 469–482

3.10 Design of Filament Wound Pressure Vessels, Chapter 29, Section VI: Design of Structures, *Polymer Composites*, Vol 1, *Structural Materials Handbook*, European Space Agency, Feb 1994

3.11 S.T. Peters, W.D. Humphrey, and R.F. Foral, *Filament Winding Composite Structure Fabrication*, SAMPE International Business Office, Covina, CA, 1999

Composite Filament Winding
S.T. Peters, editor

Copyright © 2011, ASM International®
All rights reserved.
www.asminternational.org

# CHAPTER 4

# Programming Techniques, Computer-Aided Manufacturing, and Simulation Software

A.P. Priestley

## Introduction

This chapter discusses the ways in which the evolution of filament winding software systems has capitalized on the inherent flexibility of computer numerical controlled (CNC) winding machines and enhanced their productivity. There is a detailed discussion of different types of geometries that can be wound, from the simple to the highly complex, with insight into the limitations, advantages, and challenges of each. Components covered include classic axisymmetric parts (rings, pipes, driveshafts, pipe reducers, tapered shafts, closed-end pressure vessels, and storage tanks), nonround sections (aeromasts, airfoils, box sections, and fuselage sections), curved-axis parts (elbows, ducts), and special applications (tees).

Basic winding concepts, such as band pattern, are discussed and explained, and some simple predictive formulae are introduced. Examples are given of programming various geometries using advanced software tools, and the ways in which various materials, such as rovings, towpreg, prepreg tape, and woven materials, affect winding program generation are explored.

## History—The Programming of Mechanical Machines

Before launching into the topic of advanced programming techniques and software for CNC filament winding machines, it is worthwhile to reflect on the evolution of winding machines, which has gone hand-in-hand with the evolution of the art of product realization methods.

The earliest filament winding machines were purely mechanical and focused on the manufacture of pipes. When one considers the winding of a helical path of a constant winding angle, it is purely a matter of maintaining a constant relationship between the distances the machine carriage travels per rotation of the mandrel. Any number of mechanical systems can achieve this, the usual methods being gear boxes, belts/pulleys, and chains/sprockets.

Achieving a constant pitch was the first requirement for the mechanical machine programmer. However, the product design engineer demands specific wind angles for the pipe to optimize the performance of the composite.

At this point, it is worthwhile to define *winding angle*, because this term will occur frequently in the discussion and means different things to different people. For a pipe, the winding angle is defined as the angle between the fiber direction on the mandrel surface and the axis of mandrel rotation, so, by this definition, 0° is axial and 90° is circumferential, the latter often being referred to as the hoop direction. The relationship between pitch ($P$) and winding angle ($\alpha$) is a simple one: $\tan(\alpha) = \pi\phi/P$. The astute reader will see the immediate problem, which is $\phi$, the mandrel diameter.

To achieve a specific wind angle for a particular diameter of mandrel, a specific gear ratio between carriage motion and mandrel rotation is required. Thus, the mechanical machine programmer must have available a range of interchangeable gears and a means of changing them—not the most user-friendly of programming systems! For the CNC machine, this is not a problem; it is just a matter of the program telling the mandrel how much to rotate and how far to move the carriage in the specific period of time.

The next task for the mechanical pipe winder is to set the length of the pipe. There are numerous systems that can be employed, a common one being to drive the carriage by means of a chain drive with sprockets at each end of the carriage stroke. The length is then programmed by moving the position of one of the sprockets along the carriage and then lengthening or shortening the chain.

The final programming challenge was to achieve a band pattern for winding on a mandrel.

### *Band Pattern Concept*

Band pattern is a concept that produces a systematic structure for the laying of successive bands to cover the mandrel uniformly around (axisymmetric) mandrels. A band is a flat, collimated group of fibers that have been impregnated and laid onto the mandrel. The band width of the band is related to the number of tows, the thickness of the band, and the cross-sectional area of the tows. The width of the band will usually be different before and after laydown on the mandrel. Figure 4.1 illustrates the concept of band pattern.

For a band pattern of one, successive fiber bands for successive carriage circuits of the

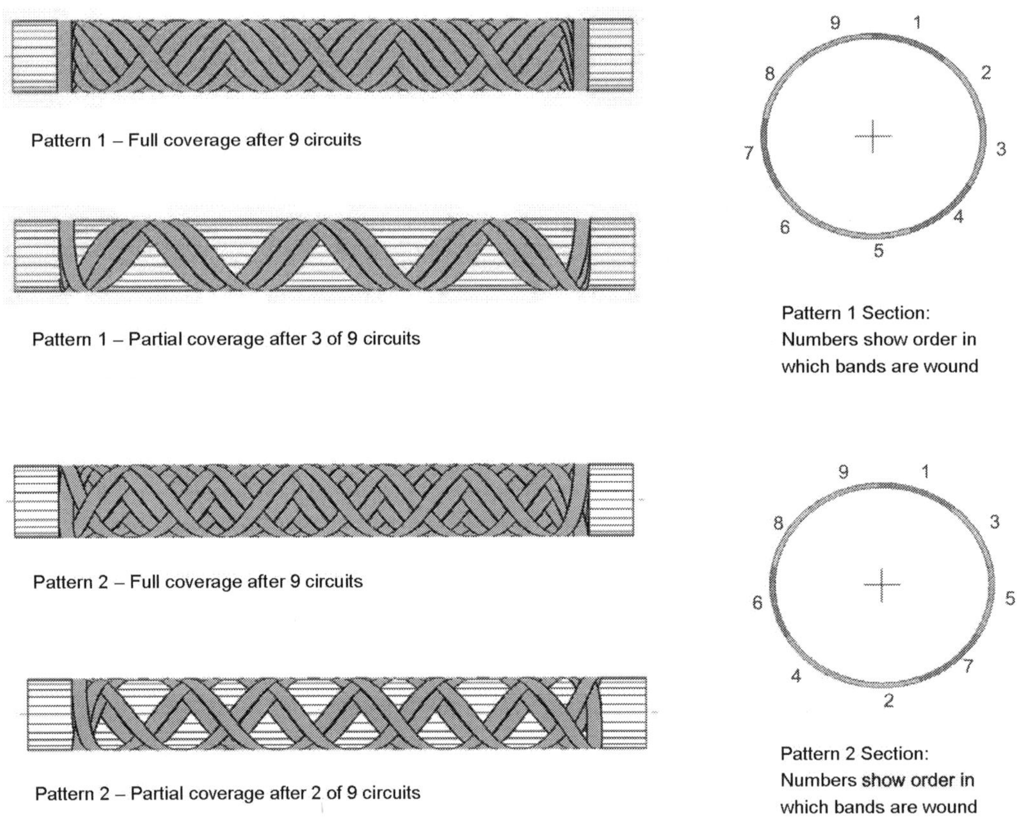


**Fig. 4.1** Development of band pattern with winding circuits

machine are laid adjacent to each other. For a band pattern of two, the bands from circuits 1, 3, 5, ... are placed adjacent to each other, as are circuits 2, 4, 6, ... So, the coverage of the mandrel develops in two swathes of material advancing around the section.

Patterns of three, four, and so on are developed in a similar manner. As the pattern number increases, the structure of the winding becomes increasingly interwoven. This has some impact on the mechanical performance of the winding. More interwoven structures tend to have a higher void content and consequently a lower fiber volume fraction. The cross-over crimp can also contribute to a reduction in mechanical performance. However, a higher interweave does give better ballistic performance and energy absorption, for example, in crash structures.

It is very unlikely for a fiber path after one circuit to return to a position exactly one band width around the mandrel from its start position. Thus, to achieve a coherent band-laying structure, the rotation of the mandrel must be modified to produce a specific rotation to make a specific band pattern.

Early mechanical machines achieved the desired band pattern by introducing some dwell at the end of the carriage stokes, where there is a specific amount of mandrel rotation while the carriage remains stationary. A typical mechanical solution was to decouple and then recouple the carriage drive with specific mechanically controlled timing.

Thus, by calculations and mechanical changes to the machine, a specific wind angle and pipe length could be achieved with a systematic pattern of laydown.

For a CNC machine, it is possible to mimic this mechanical programming approach by using some relatively simple calculations. For a number of reasons, this does not create an optimized winding program in terms of both material and time.

Referring to Fig. 4.2, position 1 shows the machine position and the fiber from machine to mandrel at the lead required for the winding angle specified. Position 2 shows the machine reaching the end of the carriage stroke. During the dwell phase, the winding angle steepens (position 3) as the fiber contact position approaches the dwell position. If the dwell rotation is long enough, the fiber will eventually reach the dwell position (position 4). However, this will be at the expense of a significant material thickening on the winding at the turning position and a significant length of winding where the fibers are not placed at the angle desired by the programmer. If the dwell is less, the machine turns around without the fiber actually reaching the dwell position. Thus, the actual length of the winding becomes a little unpredictable, and some trial-and-error winding may be required to achieve the desired finished product.

An optimized system is shown in Fig. 4.3. At position 1, there is the same lead as previously described; at position 2, the lead is shortening, driving the fiber to turn; at position 3, the lead is still reducing, and the machine overshoots the end of the winding; and at position 4, the machine is at the dwell position, and the fiber on the mandrel is also simultaneously at the same position. There can then be the minimum mandrel rotation to satisfy the band-pattern requirements before the machine continues on to the return stroke. In this way, the fiber has been turned in the minimum length of mandrel, the buildup at the dwell can be minimized, and the wound length will have been precisely controlled.

### *Geodesic Paths and Frictionally Modified Geodesic Paths*

A more advanced programming software system does not consider what the winding machine is doing but focuses on the physics of what the fiber is doing on the mandrel surface. An exactly predictable winding length can be achieved while minimizing the length or pipe

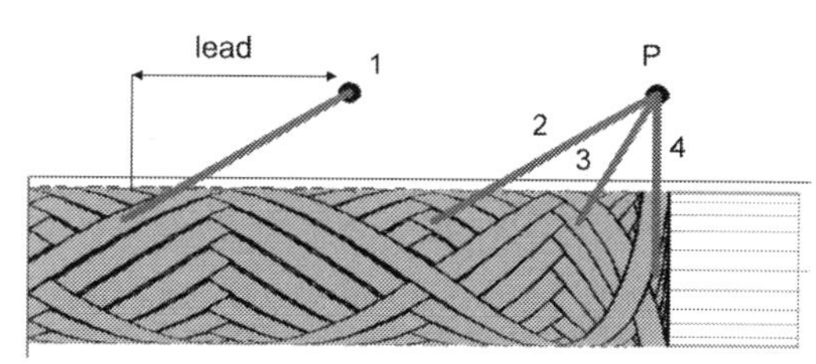


**Fig. 4.2** Simple dwell method of pipe winding

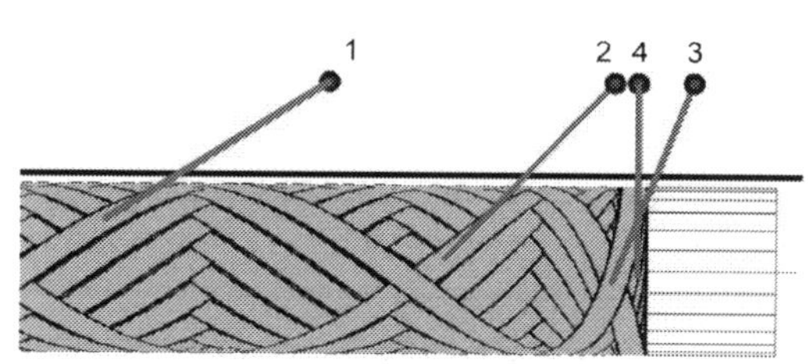


**Fig. 4.3** Advanced method of pipe winding

at the wrong winding angles and minimizing the buildup in the turning zones. This provides savings in both material and time for the manufacturer.

There are a number of principles that apply regardless of whether the component to be wound is a simple pipe or a complex nonaxisymmetric shape. The first principle is that of the geodesic path. A rigorous mathematical treatment is not necessary here, because there is a wealth of published material available for those who are interested. Suffice it to say that in simplistic terms the geodesic path represents the shortest path across the surface of the mandrel. For filament winding, this is usually a curved surface, often a bidirectionally curved surface. The principle is that the fiber material being wound is generally considered inextensible, and thus, if one considers a short portion of path that is part of a geodesic, then to deviate this portion laterally (Fig. 4.4) implies that the fiber must stretch, because being part of a geodesic, it is already on the shortest path between the points. Thus, the geodesic is generally a stable, nonslip path on the mandrel surface.

For filament winding, geodesics are a good thing. However, if only geodesic paths are employed, then even a simple pipe cannot be wound. A geodesic on a cylinder is, in fact, a helical path of constant pitch (and wind angle), so it is not possible to turn the path around at the end of the pipe unless the path deviates from the geodesic path.

Taking advantage of the principle of friction gives rise to frictionally modified geodesic forms. As the fiber progresses across the surface of the mandrel along a geodesic path, frictional resistance between the fiber and the surface offers some natural resistance to the

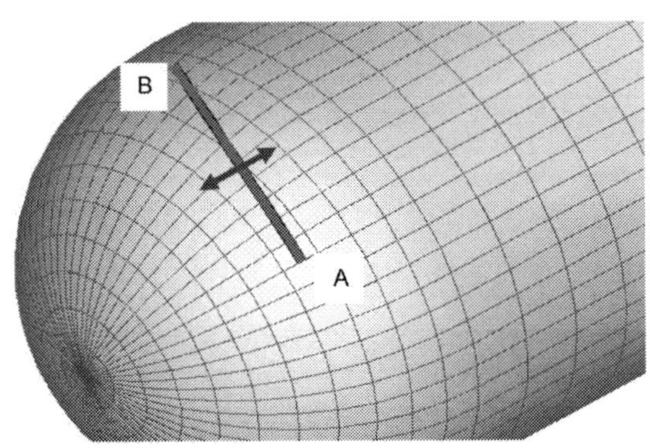


**Fig. 4.4** Geodesic path on a curved surface

fiber slipping or skidding across the surface, thus allowing some deviation from the rigidly defined path.

While friction is a useful analogy, the resistance to slippage can come from several physical mechanisms other than pure mechanical friction. For example, when winding with a tacky tow-preg (resin-preimpregnated tow), there is some physical adhesion between the fiber being applied and the mandrel. Another effect during wet winding with high fiber tension is that, at the point of fiber application, there is a tendency to compress the new fiber into the material already laid and to generate a degree of mechanical interlock.

Thus, any advanced software system used for programming filament winding machines must incorporate the concept of some friction, and the user must be able to specify a limiting (i.e., nonslip) value for the materials being used.

## Axisymmetric Shapes

The vast majority of geometries for winding are axisymmetric. An axisymmetric body, as the name implies, is a body with an axis of symmetry, a body of revolution. Cylinders, tapered shafts, spheres, and cones are examples of axisymmetric shapes. The most common manufactured parts are pipes for fluid transport or cylinders with closed ends (usually spherical, elliptical, or torispherical) for pressure or storage vessels.

Filament winding has a number of advantages over other manufacturing processes, including being one of the few automated processes available that is capable of delivering excellent mechanical properties from continuous aligned fibers. The main winding process limitations are the geometric constraints on being able to productively wind general shapes at any winding angle. Axisymmetric bodies lend themselves to the filament winding process for a number of reasons.

Because an axisymmetric body is rotationally symmetric, it is only necessary to program a single circuit of the fiber wind path (Fig. 4.5). As long as the rotation of this circuit is such that it makes a repeatable band pattern (as discussed earlier in this section), the circuit can be repeated to provide a fiber layer completely covering the mandrel. A nonaxisymmetric shape does not have such a simple strategy for covering the mandrel, and, in general, the complete

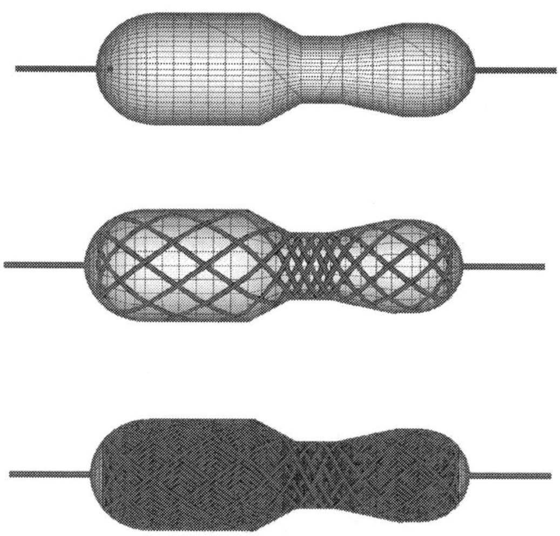

**Fig. 4.5** Single circuit repeated for axisymmetric coverage

coverage path must be calculated, with each circuit being unique.

For a simple axisymmetric case such as a pipe, the number of circuits required to cover a pipe is determined by considering a fiber band of true width, $B$, at a winding (helix) angle of $\alpha$. By applying some simple trigonometry, the width of the band in the circumferential direction is $B/\cos(\alpha)$, and dividing this into the pipe circumference, the number of circuits ($N$) required to cover a pipe of diameter $D$ is given by $N = \pi \cdot D \cdot \cos(\alpha)/B$.

The number of circuits must, by necessity, be a whole number, so if the value is rounded up, there will be some overlap of the band edges, and if the value is rounded down, there will be gaps between each of the bands. Winding with overlaps will give some increased unevenness of the outer surface of the winding, while gaps will tend to fill with resin and may result in some decrease in the final fiber volume fraction. Neither may actually be a problem, but there is always the possibility to eliminate this effect by adjusting the actual band width by changes in the fiber-delivery system, for example, changing the number or spacing of the guides for the fiber tows or by making a small adjustment to the actual winding angle to produce a whole number of circuits.

For a more general axisymmetric shape, the $N$ value should be calculated for all positions along the geodesic path and the largest value used if full coverage is required. In practical terms, this limiting case is almost always at the largest cross section of the mandrel.

### The Clairault Relation

There are other interesting properties of axisymmetric bodies, as discovered by the prominent French mathematician Alexis Claude de Clairault. Clairault's relation (Ref 4.1) is a formula in classical differential geometry. The relation is valid for any point on a geodesic path on an arbitrary surface of revolution and gives:

$$r \cdot \sin(\alpha) = \text{Constant} \qquad \text{(Eq 4.1)}$$

where $r$ is the radial distance of any point on a geodesic path from the axis of revolution, and $\alpha$ is the angle between the tangent vector and the latitudinal circle, or, to nonmathematicians, $\alpha$ is the winding angle.

This relationship leads to some fundamental principles incorporated into most filament winding processes. It states that along any geodesic path, as the radius of the body decreases, the winding angle increases, and conversely, as the radius increases, the winding angle decreases.

For example, consider a cylinder of 12 in. (300 mm) diameter (6 in., or 150 mm, radius) with hemispherical dome ends (a typical pressure vessel). If the winding angle on the cylindrical section of the vessel is 20°, then $150 \cdot \sin(20) = 51.3$ = constant. For a point on the dome where the radius is 4 in. (100 mm), then $100 \cdot \sin(\alpha) = 51.3$, and $\alpha$, the winding angle, is 30.9° at this location. Thus, the winding angle is increasing as the radius of the dome is decreasing. Now consider the special position when the path is at a 90° winding angle. At 90°, the path is travelling in the circumferential direction and is the turnaround position of the path. Because $\sin(90) = 1$, the Clairault relationship relates that $r \cdot \sin(90) = 51.3$, and thus, $r = 2$ in. (51.3 mm) at the turning position. That is, the end-opening diameter is 4 in. (102.6 mm).

Note that winding geodesically at 20° on a cylinder of this diameter will always produce this size of end opening, regardless of the actual shape of the end dome itself. There was one simplification in this discussion: The roving or tape being used in winding will have a finite width, often referred to as the band width. When determining the trajectory on the mandrel surface of the center of the band, the end-opening diameter calculated previously will, in actuality, be smaller by a full band width (i.e., half a band at each side of the opening).

### *Limitation of Antislip Paths*

So far, the possibility of using friction, the resistance to slipping, to modify geodesic paths has been discussed. In the previous example, the geodesic end-opening radius was 2 in. (51.3 mm). Using a value for friction, $\mu = 0.15$ (friction is dimensionless), which is a value that may typically be seen in normal winding, and making a quick trial using Cadfil (Crescent Consultants Ltd.) (Ref 4.2), a filament winding software, it is found that openings (neglecting band width) of between 3.75 and 0.3 in. (95.9 and 7.3 mm) are possible for this geometry and wind angle. What can be seen here is that for this wind angle, it is not possible to avoid winding onto the lower part of the dome; conversely, with higher wind angles, it may not be possible to close the end of the vessel to a specified diameter, such as for an end boss.

Frequently, when such a limitation is first encountered, the initial instinct is to think there is some deficiency of the winding machine or the programming software or both. It must be firmly kept in mind that no software system is capable of changing the laws of physics!

### *Fiber Bridging*

A further restriction for winding is that of fiber bridging, where winding onto a concave shape is attempted and, as a result, the fibers do not remain in contact with the surface. Axisymmetric shapes have the advantage that, in the circumferential direction, there is always positive curvature; this can be seen in the rocket casing shown in Fig. 4.6. This shape is axisymmetric, but for the fiber to remain in contact with the surface in the concave zone, the winding angle in the area must be almost circumferential. The effect of fiber bridging at a lower angle can be seen in Fig. 4.7, as developed from a Cadfil software winding simulation. For this type of geometry, the need to develop nonslip paths and to avoid fiber bridging can quite severely restrict the designer in what fiber directions can be achieved. In this instance, the designer must modify the wound shape to fit the constraints of the winding process, for example, by blending out the concave area, or apply some material at more favorable orientations by some other method, such as the local hand placement of fabric reinforcements.

### *Thickness Distribution for Axisymmetric Shapes*

A further challenge where product and process design for the winding interact is the consideration of the thickness of the winding. For axisymmetric shapes, the way a single circuit can be replicated to provide full coverage around a mandrel has been discussed. When

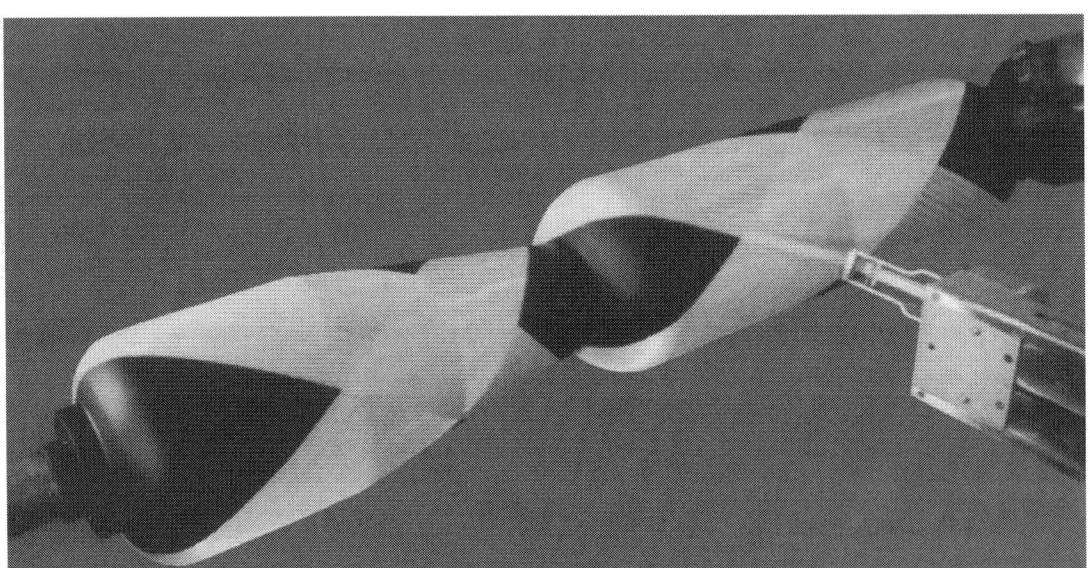

**Fig. 4.6** Half-wound rocket casing showing constant positive curvature in the circumferential direction. Courtesy of Crescent Consultants Ltd. in collaboration with Pultrex Ltd.

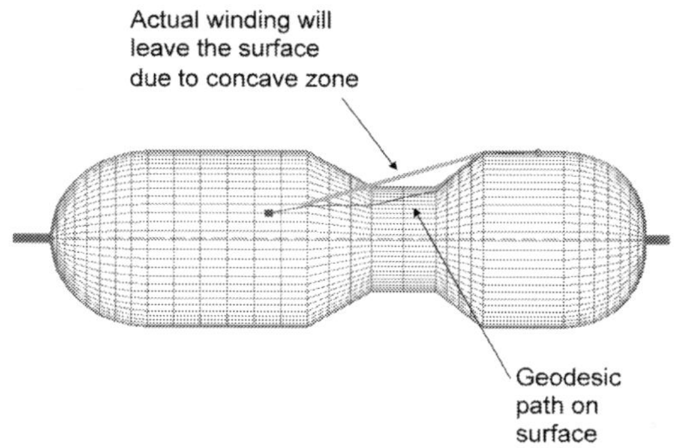


**Fig. 4.7** Fiber bridging illustration

one considers that the same number of bands or rovings are being placed at all positions on the circuit, it is intuitively obvious that as the mandrel diameter increases or decreases, the thickness increases or decreases proportionately as a simple matter of conservation of mass. There is another effect at play, however. As the winding angle increases, the band becomes more oblique to the section of the mandrel, and the effective thickness increases in proportion to $1/\cos(\alpha)$. This is most obvious when winding low angles on a closed-end pressure vessel, where, at the polar end openings, the winding is several times thicker than on the body of the cylinder.

## Software Systems for Numerical Control Winders

Some of the physical considerations that go into controlling a filament winding machine have been discussed. Simple parts can be programmed by hand calculation or a spreadsheet, if the programmer is so inclined and possesses sufficient knowledge and expertise. For more complicated parts, and especially for machines with more than three axes of coordinated motion, some form of software is essential to be able to economically generate quality winding paths. Many simple software systems are available to replicate hand calculations or specific types of winding that can be described using analytical equations, such as for winding spheres.

### *Brief History of Early Winding Software Theory*

The concept of using computers for calculating repeating geodesic paths on axisymmetric shapes was identified as early as 1966 by Dunbar (Ref 4.3). His method could be carried out by purely graphical techniques, but the speed of calculation was greatly increased through the use of a computer.

A system for calculating both geodesic paths and constant resistance to slippage paths was reported by Menges, Wodicka, and Barking (Ref 4.4) in 1978. This was limited to surfaces of revolution and uses the Clairault relationship condition to predict geodesic paths. This is discussed in detail by Lyusternik (Ref 4.5).

By 1980, Hille and Menges (Ref 4.6) were advocating a practical teach-in system for complex mandrel shapes. They had found that "for rotationally symmetrical cores, the control data can be determined by computer programs, but because of the scope involved, these can generally only be run on large computers." For more complex geometries, they could not see any possibility of producing winding patterns economically by computer.

In 1983, Braun (Ref 4.7) provided good insight to the problems of computer-aided design and computer-aided manufacturing (CAD/CAM) for filament winding. He explained many of the problems involved, discussed the use of conical sections, triangular patches, and so on to obtain solutions, and suggested that a CAD/CAM system was theoretically possible.

Bowen (Ref 4.8) discussed the various methods of fiber-track prediction available in 1984, along with the possible use of thermoplastic preimpregnated fiber tapes to remove some of the fiber slippage restrictions. He also made the point that although it is possible to develop a CAD/CAM system "for the production of complex composite structures on a fully automated basis," this was unlikely in the near future.

In November 1985, Eckold and Wells (Ref 4.9) reported a CAD/CAM system that could predict geodesic paths on three-dimensional (3-D) surfaces. This was based on solution of the geodesic differential equations.

### *Advanced Winding Software*

Filament winding and computer technology have advanced considerably from the early references discussed previously. The calculation of nonslip geodesic and modified geodesic paths, whether by solving geodesic differential equations or a discretized system on surface patches for which an analytical solution exists, is well documented and established (Ref 4.7, 4.9). Filament winding has moved from the domain of the academic researcher into established practice for composites manufacturing. Manufacturing companies are primarily interested in ready-made software solutions that are affordable, easy to use, and reliable.

There are four elemental stages of any advanced software system, although some of the stages may be rolled into one operation as far as the user is concerned:

- *Mandrel definition:* Creating or importing from a CAD package a 3-D representation of the surface of the mandrel that is to be wound
- *Path generation:* Creating feasible paths on the defined surface of the mandrel. The paths must be stable trajectories that do not exhibit slippage or significant fiber bridging.
- *3-D machine path generation:* Generation of a path in free space around the mandrel so that the machine payout eye will follow to place the fiber path correctly on the mandrel surface. Filament winding paths are unlike cutter paths that are generated by CNC machining systems because the tool on the filament winding machine is the fiber-delivery (payout) head, and this does not actually come in contact with the mandrel. This machine position is in free space somewhere around the mandrel.
- *Postprocessing:* This step converts the 3-D machine positions into data that the winding machine controller can interpret. The postprocessing software must be configured for the number and type of axes available on the winding machine, the relationship between mandrel and machine units of measure, and the syntax and format of the data that the winding machine is capable of interpreting.

**Mandrel Geometry.** There are different approaches to creating mandrel geometry for winding program generation. This section covers the approaches used in the Cadfil software, because this is the system with which the author is most familiar, although this approach is typical of the strategies employed by other software applications as well. For common geometries, parametric mandrel definition makes a great deal of sense because it is both quick and simple. For example, a pipe is characterized by only a length and diameter, while for a standard pressure vessel, there is a diameter and length for the cylinder as well as some parameters for the type of vessel end cap, examples of which can be seen in Fig. 4.8. In Cadfil, a similar approach is used for pipe bends (elbows) and pipe tees, because these generic shapes are geometrically easy for a user to define in this way.

For more general axisymmetric shapes, some form of shape-editing tool is required. However, because an axisymmetric body is a body of revolution, the surface can be defined purely by using a longitudinal cross section. Thus, only two-dimensional (2-D) geometry is required; in Cadfil, this section is defined by entering a list

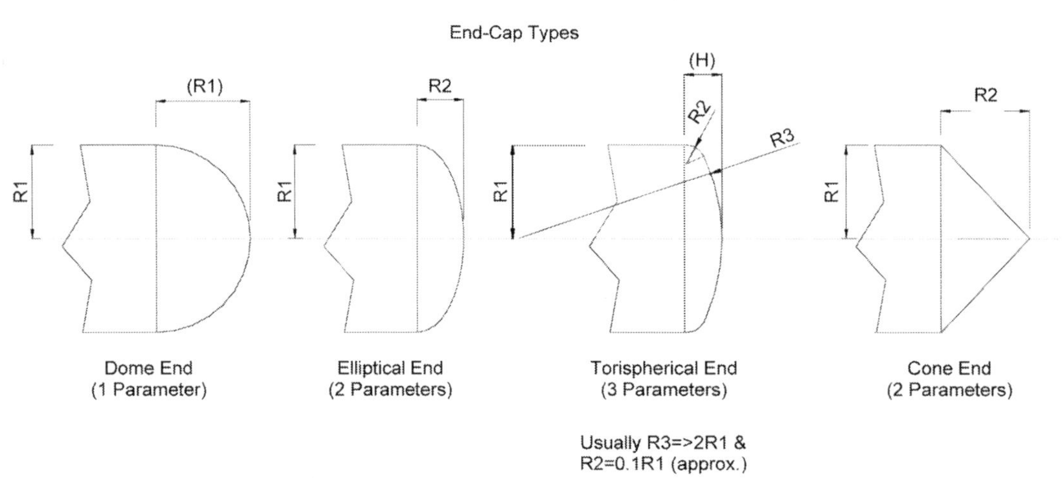


**Fig. 4.8** Types of end-cap geometries

of 2-D data points that are joined by lines or curve segments of specified radius.

The final way of mandrel definition is to import data from a CAD program; the most common format supported in filament winding software packages is the drawing exchange format (DXF). The DXF format is also the most widely supported CAD format for 2-D data.

**Surface Path Generation.** The generation of nonslip paths is the core of any filament winding software system. Regardless of the methodology of the mathematical engine that generates the paths, there are two approaches to a user interface. The first approach is parameter driven, which puts some constraints on the winding path. Thus, for the example of the ubiquitous closed-end pressure vessel, the user may specify that a winding layer must touch boss diameters on each end cap and may also specify a target winding angle along the cylinder. Then, given some frictional limit, the winding path can be automatically calculated. This approach is very good for standard geometries where the requirements are easily defined, but a more flexible interactive approach is required in instances of complex shapes. The second, more flexible approach is to allow the user to fix an initial point of the winding paths, specifying a position and direction (winding angle), and then for the user to "drive" the path across the mandrel surface while monitoring the position and angle. The user can then interactively modify the applicable friction value (within defined limits) to steer the path to achieve the desired outcome. This approach requires a little more thought and effort on the part of the user but ultimately allows the designer more creativity.

**3-D Machine Path—Payout Path Strategies.** This is a simple process, because the fiber under tension is placed onto the mandrel in a straight line tangent to the surface path at the point of contact. There are only two methods that seem to be used; the simplest of these is the constant free fiber length method, where the programmed machine position is a specified distance along the tangential vector from the surface contact point. This is simple insofar as the input data are minimal; however, there are drawbacks. First, if the winding angle is very low (axial), the path diverges away from the mandrel surface very slowly. So, for a given free fiber length, the path can be too close to the mandrel surface in this case and then, at higher wind angles, it can be too far away, leading to excessive and wasteful machine motion. Second, for odd-shaped mandrels (tee joints being a good example), a tangent line from the surface on the cross piece of the tee may actually intersect with the branch of the tee, causing a machine/mandrel crash.

The other approach is to define a control surface or enveloping surface around the mandrel and have the constraint that the programmed machine position will lay on these surfaces. It is then simply a case of following the tangential direction from the mandrel contact point until the first interaction with the enveloping surface. As long as the clearance surfaces are suitably defined, clash-free winding is guaranteed, as well as efficient use of the winding machine by minimizing the amount of machine motion.

The control surface envelope can be created and specified using exactly the same software techniques that are used for specifying the mandrel geometry. Indeed, with parametric programming, a scenario can simply be generated using clearance values from the mandrel surface.

**Postprocessing.** Writing software that converts the 3-D machine position data into data that the winding machine controller can understand is a relatively simple task in the case of one winding machine and one CNC controller. For commercial software, it is both advantageous and essential to consider all the different machine configurations, from a simple two-axis machine to one with six or more degrees of freedom. Indeed, postprocessing to allow winding by using a standard commercially available six-axis robot is also desirable. When the wealth of different types of CNC controller systems that are available and used by machine builders is also considered, the problem multiplies. The solution to this problem requires some sophistication in allowing the user or machine builder to configure the postprocessor for the machine at hand. If this is done well, the user does not have to worry about the underlying configuration and can focus on the task of selecting which of the available machine control strategies is best to employ.

### *Enhanced Programming Software*

The previous sections discussed the building blocks required for an advanced software system for filament winding. Most organizations that have produced such software have added various enhancements. Such enhancements include 3-D visualization of the wound structure and graphical display of the thickness distribution on the surface. The 3-D simulation of the motion of the machine around the mandrel is now also possible on low-cost computer hardware, using well-established 3-D graphical techniques and algorithms.

Another significant enhancement for some users is the ability to export 3-D data for shape, fiber direction, and winding thickness for use in analysis systems such as finite-element analysis. For axisymmetric systems, it is often more useful and flexible to export 2-D thickness, shape, and angle data relating to a longitudinal section, because the analysis may often have to consider modeling features other than the winding in the final component assembly.

## Nonaxisymmetric Components

Rotationally symmetric components represent the vast majority of components made by filament winding. However, nonaxisymmetric components represent additional challenges, and indeed, many geometries are totally unsuited to manufacture by filament winding. The nonaxisymmetric components that are commonly wound are prismatic nonround sections and pipe fittings such as elbows and tee joints.

The principles of making a surface representation of the mandrel and generating a geodesic or modified geodesic path on this nonaxisymmetric surface still apply. The difficulty is having a strategy to make coverage of the surface in an organized manner with fiber directions that have relevance to the requirements of the component.

In general, each path or circuit on a nonaxisymmetric component will be unique, and the technique of using the rotational symmetry to cover a mandrel by repeating a single circuit cannot be applied. The result is that the volume of CNC data describing such a part will be significantly larger than for an axisymmetric part of approximately the same size. This used to be a problem for the capabilities of CNC controllers; however, advances in controller design have mitigated this problem because sufficient memory and storage can be provided for an economical cost.

A universal software system that can cope with winding any geometry is not really a practical reality, because different classes of geometry require different approaches. Software available in the marketplace is targeted to specific categories and groups of geometries, some of which are discussed in the following sections.

### *Nonround Prismatic Sections*

An example of such geometry is a tube with a hexagonal bore. Such a shape is not so dissimilar to a round tube and indeed can be wound by standard axisymmetric methods. A good assumption is to calculate the perimeter of the cross section and assume a pipe diameter that gives the same perimeter. This works most satisfactorily for higher-angle, open-ended winding and gives reasonable results for lower-angle winding using a modified dome end.

A square-section mandrel can also be wound using such axisymmetric methods but is not as satisfactory as a hexagon. When winding on a square-section mandrel, there is winding angle variation between the corners and the flats due to the variation in the diameter of the part at these respective points. There is also some variation in thickness around the section for the same reason. However, for many applications, these drawbacks may not present a problem.

For rectangular sections, as the shape becomes less square (more eccentric), the approximation to a round structure and the quality of winding becomes increasingly less successful, and finding a solution for low-angle winding is much more difficult. In this instance, a nonaxisymmetric programming methodology is required.

For prismatic shapes, a constant wind angle (geodesic) path along the length of the part is quite easy to calculate, and calculating parallel geodesic paths separated by a band width is also relatively simple. The only technical problem is to link all these paths to make a continuous winding.

For winding such a shape with open ends, this joining process is very similar to winding a pipe. There is a turning zone at each end of the part, in which the wind angle gradually steepens until the winding is circumferential. There is then some dwell, and the path turns back down the mandrel, picking up on a return path at the end of the turning zone. Because the turning and/or dwell need adjustment such that the mandrel is covered in a systematic way, each linking path is liable to be unique.

If the component requires a very low angle, winding the turning zone would be very long and would not be economical. A different method of linking the paths will be required.

### Winding a Nonround Aero Section Boat Mast

An example of low-angle winding on a nonround prismatic section can be seen in Fig. 4.9 and 4.10. This component, over 98 ft (30 m) in length, required mostly low-angle winding (6 to 10°). Fortunately, the ends of the part were not required, so the technique employed having a flat end plate on the mandrel with a pin ring. This allowed the paths to exit off the end of the mandrel, wrap onto the mandrel support shaft, and then pick up on the correct position upon re-entering onto the mandrel. The pin ring prevented slippage due to any angle discontinuity at the entry and exit points.

### Nonround Road Tanker

The final example in this section is winding of a road transportation tanker with nonround cross section; this can be seen in Fig. 4.11. This component required a mixture of hoop and low-angle winding, and, more challengingly, the low-angle winding over the end cap was required in the finished structure. For this, the end paths must pass over the end caps within the limits of friction and pick up on a suitable return path. Because the section is constantly changing, there is a degree of incompatibility with the entry and resultant exit angles, and thus, the mandrel cross-sectional shape, end dome shape, and winding pattern and angle must be fully considered in the component design to ensure that the winding is fully feasible. Again, each end-cap path is unique.

### Pipe Elbows and Tee Winding

There are two types of winding commonly used on elbows; the first is hoop winding, where the mandrel is covered in a single circuit. This usually occurs when the winding is being used to reinforce a pipe to increase its pressure rating or to provide environmental protection. This method is also used with fabric tapes to provide a bidirectional fiber architecture. The second type of winding is helical winding with a ± angle-ply layup, where the mandrel is covered by a number of circuits. A turning zone at the ends of the bend or on a parallel extension is the

**Fig. 4.9** Nonround, prismatic Aero section boat mast. Courtesy of Crescent Consultants Ltd. in collaboration with Pultrex Ltd.

normal methodology. This is exactly analogous to helical winding of an open-ended pipe, the difference being that the pipe is curved. As such, the bands on the inside of the bend are closer together, and those on the outside of the bend are farther apart. It is normal to fix the band width for the outside of the bend to give full coverage, while band overlap gives increased thickness on the inside of the bend. It must also be remembered that each circuit is unique.

It is not possible to fully cover an elbow with low-angle winding because the fibers bridge across the inside of the bend. The lowest angle that can be used for full coverage is a function of the bend curvature in relation to the pipe diameter. For best results, a curvature of greater than 2.5 diameters gives the best results, and full coverage with any winding angle above 35° should be possible.

Pipe tees do not naturally lend themselves to winding, because the fiber orientations are very restricted by the geometry. The most practical geometry is the 90° equal tee with blend radii of $D/2$, where $D$ is the pipe diameter. There are a number of standard winding patterns that can be wound on such a tee. A diagrammatic description of these can be found in U.S. patent 4,601,770 (Ref 4.10). The best route the author has found for winding these parts is to generate

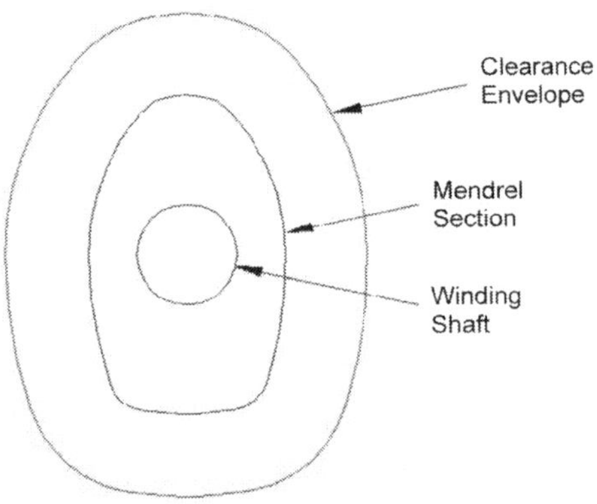


**Fig. 4.10** Computer-aided design section drawing of an Aero boat mast cross section

**Fig. 4.11** Road tanker winding. Courtesy of Crescent Consultants Ltd. in collaboration with CNC Technics Pvt. Ltd.

single-circuit winding patterns using geodesic surface modeling and to then replicate these using parametric methods with friction-limit checking. A complete coverage is built up by using a number of these patterns.

## REFERENCES


4.1 M. do Carmo, *Differential Geometry of Curves and Surfaces*, Prentice Hall, 1976, p 257

4.2 Cadfil, Crescent Consultants Ltd., United Kingdom, http://www.cadfil.com

4.3 G.R. Dunbar, "Filament Winding," Paper 66-MD-86, ASME Meeting, May 9–12, 1966

4.4 G. Menges, B. Wodicka, and H.L. Barking, "Non-Geodesic Winding on a Surface of Revolution," 33rd Tech. Conference, Reinforced Plastics/Composites Inst., SPI Inc., 1978

4.5 L.A. Lyusternik, Shortest Paths—Variational Problems, *Popular Lectures in Mathematics*, Vol 13, Pergamon Press Ltd., 1964

4.6 E.A. Hille and S. Menges, GRP Pipe Fittings Produced on a Computer Controlled Filament Winding Machine, 35th Tech. Conference, Reinforced Plastics/Composites Institute, SPI Inc., 1980

4.7 W. Braun, Problems of Determining and Preparing Control Data for Non-Rotationally Symmetrical Winding Bodies Explained Using the Geodetically Wound Elbow, Paper 17, *Proceedings Third Technical Conference SAMPE European Chapter, Engineering with Composites*, Vol 2, 1983

4.8 D.H. Bowen, "Filament Winding in the 1980s," Fibre Reinforced Composites '84 Conf., April 3–5, 1984 (University of Liverpool)

4.9 G.C. Eckold and G.M. Wells, Computer Aided Design and Manufacture of Advanced Composites, *Proceedings of Second Conference on Materials Engineering, I. Mech. E. & P.R.I.*, Nov 5–7, 1985

4.10 R. Ulrich and R. Friedrich, Fiber Reinforced Plastic Pipe Tee, U.S. Patent 4,601,770, Sept 19, 1983

Composite Filament Winding
S.T. Peters, editor

Copyright © 2011, ASM International®
All rights reserved.
www.asminternational.org

# CHAPTER 5

# Fiber Trajectories on Shells of Revolution—An Engineering Approach

Sotiris Koussios, Delft University of Technology

## Introduction

This chapter outlines a method for mathematically describing fiber trajectories on a shell of revolution. Both geodesics and nongeodesics are considered, combined in a single theory. There are a considerable number of excellent textbooks regarding these issues, but the theory is usually presented in a rather rigorous way and cannot directly be put into practice. All derivations in this chapter are based on simple intuitive geometric observations and simple vector calculus; at the end, the only input needed is the parametric description of the meridian profile.

After a short outline of the basic geometry of shells of revolution, the focus shifts to fiber trajectories and their characteristic metrics, angles, and vectors. Next, emphasis is given to the determination of various kinds of curvatures that eventually lead to the derivation of (non-) geodesic fiber trajectories according to a predetermined coefficient of friction. The chapter ends with the analysis of nongeodesics on conical segments, annuli, and cylinders. Besides the existence of analytical solutions, conical segments are chosen because they represent an easy way to discretize arbitrary shells of revolution.

## Geometry of a Shell of Revolution

The basic mechanism to create a shell of revolution is to revolve a certain profile (meridian profile) around an axis over 360° (Fig. 5.1). The most generic way to define a meridian profile, **M**, is a parametric description:

$$\mathbf{M} = \{\rho(t), \zeta(t)\}, \quad t_0 \le t \le t_1 \qquad \text{(Eq 5.1)}$$

The parameter $t$ can, for example, be the length along the meridian curve or a goniometric quantity; for every $t$, a radial ($\rho$) and an axial ($\zeta$, height) coordinate are obtained (Fig. 5.1). The meridian profile is then revolved around the $\zeta$-axis. The corresponding mathematical definition is:

$$\mathbf{S}(t,\phi) = \begin{Bmatrix} x(t,\phi) \\ y(t,\phi) \\ z(t,\phi) \end{Bmatrix} = \begin{Bmatrix} \rho(t)\cos\phi \\ \rho(t)\sin\phi \\ \zeta(t) \end{Bmatrix}, \quad \begin{matrix} t_0 \le t \le t_1 \\ 0 \le \phi \le 2\pi \end{matrix} \qquad \text{(Eq 5.2)}$$

On a shell of revolution, two main directions are distinguished, which are vertical to each

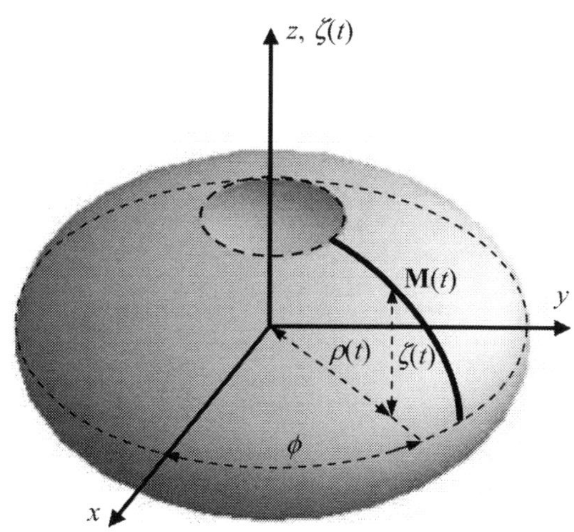


Fig. 5.1 Definition of a shell of revolution

other. The first, meridional direction (north ↔ south), is characterized by a fixed $\phi$ value. The second, parallel direction (east ↔ west), is governed by a constant $t$ value.

The fiber bundle (dotted line, $\mathbf{C}(t)$) is now placed under an angle $\alpha$ with respect to the meridian (gray line) (Fig. 5.2). For determining the kinematic solution of the associated winding process, the required information about a point on that fiber bundle consists of two parts: the position of that point and the orientation of the fiber bundle at that point. Because the fiber is placed under an angle $\alpha$ with respect to the meridian, its orientation $\mathbf{t}$ will be a vector combination of the parallel (vector $\mathbf{B}$) and meridional ($\mathbf{T}$) directions (also referred to as main directions of the shell). Hence, the two main directions must be implemented first. When following the meridian direction, the (unit-length) orientation vector is given by:

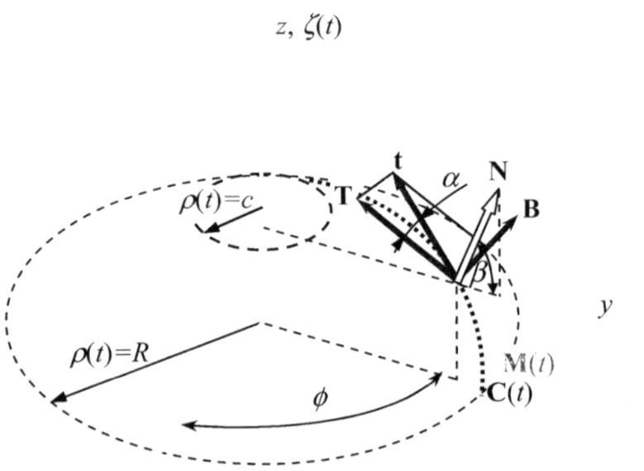


**Fig. 5.2** Shell-related orientation vectors (capital letters) and tangent vector **t** of a fiber path placed on that shell

$$\mathbf{T}(t,\phi)=\frac{\mathbf{S}_t(t,\phi)}{\underbrace{\sqrt{\mathbf{S}_t(t,\phi)\cdot\mathbf{S}_t(t,\phi)}}_{\sqrt{G}}}=\frac{1}{\sqrt{\rho'^2(t)+\zeta'^2(t)}}\begin{Bmatrix}\rho'(t)\cos\phi\\ \rho'(t)\sin\phi\\ \zeta'(t)\end{Bmatrix}\qquad\text{(Eq 5.3)}$$

where the square root term provides the connection between the meridional contour length increment, $ds$, and the increment $dt$ of the parameter $t$ (Fig. 5.2, 5.3):

$$|ds|=\sqrt{(d\rho)^2+(d\zeta)^2}=\sqrt{(\rho'dt)^2+(\zeta'dt)^2}=\left(\sqrt{\rho'^2(t)+\zeta'^2(t)}\right)dt=\sqrt{G}dt\qquad\text{(Eq 5.4)}$$

In a similar fashion, a walk around the periphery (east ↔ west, fixed $t$ value) gives the following unit-length orientation vector $\mathbf{B}$:

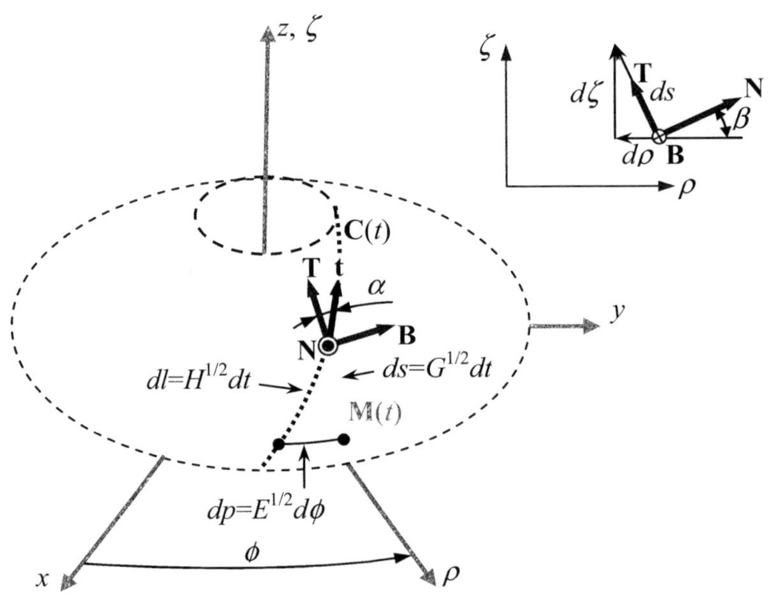


**Fig. 5.3** Metrics of a fiber trajectory on a shell of revolution

$$\mathbf{B}(t,\phi)=\frac{\mathbf{S}_\phi(t,\phi)}{\underbrace{\sqrt{\mathbf{S}_\phi(t,\phi)\cdot\mathbf{S}_\phi(t,\phi)}}_{\sqrt{E}}}=\frac{1}{\sqrt{\rho^2(t)}}\begin{Bmatrix}-\rho(t)\sin\phi\\ \rho(t)\cos\phi\\ 0\end{Bmatrix}=\operatorname{sgn}(\rho(t))\begin{Bmatrix}-\sin\phi\\ \cos\phi\\ 0\end{Bmatrix}\qquad\text{(Eq 5.5)}$$

where sgn(#) stands for the sign of $\rho(t)$. Although the sign is always positive, this rigorous definition must be maintained for later derivations. The square root term relates $dt$ to the length increment $dp$ in parallel direction (Fig. 5.2, 5.3):

$$dp=|\rho|d\phi=\sqrt{E}d\phi\qquad\text{(Eq 5.6)}$$

For a shell of revolution, the vectors **T** and **B** are perpendicular to each other. This implies that the two main shell directions are linearly independent:

$$\underbrace{\mathbf{S}_t(t,\phi)\cdot\mathbf{S}_\phi(t,\phi)}_{F}=0\qquad\text{(Eq 5.7)}$$

This result can immediately be verified by multiplying Eq 5.3 by Eq 5.5. The coefficients $\{E, F, G\}$ are referred to as coefficients of the first fundamental form (Ref 5.1–5.3). Perpendicular to the **B**-**T** plane (Fig. 5.2), the vector **N** expresses the orientation of the surface at a point $\{t, \phi\}$:

$$\mathbf{N}(t,\phi)=\mathbf{B}(t,\phi)\times\mathbf{T}(t,\phi)=\frac{\mathbf{S}_\phi(t,\phi)\times\mathbf{S}_t(t,\phi)}{\sqrt{E}\sqrt{G}}=$$

$$=\frac{1}{\sqrt{\rho^2(t)}}\frac{1}{\sqrt{\rho'^2(t)+\zeta'^2(t)}}\begin{Bmatrix}\rho(t)\zeta'(t)\cos\phi\\ \rho(t)\zeta'(t)\sin\phi\\ -\rho(t)\rho'(t)\end{Bmatrix}$$

$$=\frac{\operatorname{sgn}(\rho(t))}{\sqrt{\rho'^2(t)+\zeta'^2(t)}}\begin{Bmatrix}\zeta'(t)\cos\phi\\ \zeta'(t)\sin\phi\\ -\rho'(t)\end{Bmatrix}\qquad\text{(Eq 5.8)}$$

Because this vector originates from the cross product of two unit-length vectors, its length is also equal to 1.

## Fiber Trajectories on a Shell of Revolution

Returning to Fig. 5.2, one can immediately observe that the orientation vector **t** of a fiber path (dotted line), placed on the mandrel **S**, can be written as follows:

$$\mathbf{t}=\mathbf{T}\cos\alpha+\mathbf{B}\sin\alpha\qquad\text{(Eq 5.9)}$$

Equations 5.3 and 5.5 provide:

$$\mathbf{t}=\begin{Bmatrix}-\operatorname{sgn}(\rho(t))\sin\alpha\sin\phi+\dfrac{\rho'(t)}{\sqrt{G(t)}}\cos\alpha\cos\phi\\ \operatorname{sgn}(\rho(t))\sin\alpha\cos\phi+\dfrac{\rho'(t)}{\sqrt{G(t)}}\cos\alpha\sin\phi\\ \dfrac{\zeta'(t)}{\sqrt{G(t)}}\cos\alpha\end{Bmatrix}\qquad\text{(Eq 5.10)}$$

The length of the obtained vector is equal to 1 ($\mathbf{t}\cdot\mathbf{t}=1$). In a more general consideration, the definition of a surface relies on two independent coordinates, for example, $\{t, \phi\}$. Linking these coordinates to each other by means of a function, for example, $\{t, \phi(t)\}$, will result in a curve in space that depends on one variable only ($t$). Therefore, a fiber bundle, placed on the surface of a shell of revolution, may be governed by:

$$\mathbf{C}(t)=\begin{Bmatrix}\rho(t)\cos(\phi(t))\\ \rho(t)\sin(\phi(t))\\ \zeta(t)\end{Bmatrix},\qquad t_0\le t\le t_1\qquad\text{(Eq 5.11)}$$

The unit orientation vector tangent to this curve can be obtained by differentiation of $\mathbf{C}(t)$:

$$\frac{\mathbf{C}'(t)}{\sqrt{\mathbf{C}'(t)\cdot\mathbf{C}'(t)}}=\frac{1}{\underbrace{\sqrt{G(t)+E(t)\phi'^2(t)}}_{\sqrt{H}}}\begin{Bmatrix}\rho'(t)\cos(\phi(t))-\rho(t)\phi'(t)\sin(\phi(t))\\ \rho'(t)\sin(\phi(t))+\rho(t)\phi'(t)\cos(\phi(t))\\ \zeta'(t)\end{Bmatrix}\qquad\text{(Eq 5.12)}$$

The square root argument in the denominator of this vector has a special meaning (Fig. 5.3); it represents the fiber length increment $dl$:

$$dl=\sqrt{(ds)^2+(dp)^2}=\sqrt{G(dt)^2+E(d\phi)^2}=$$
$$=\sqrt{G(dt)^2+E(\phi'dt)^2}=\sqrt{G+E\phi'^2}dt=\sqrt{H}dt\qquad\text{(Eq 5.13)}$$

where Eq 5.4 and 5.6 have been substituted. By comparing Eq 5.10 with 5.12, it can be concluded that (Fig. 5.3):

$$\cos\alpha = \frac{ds}{dl} = \sqrt{\frac{G}{H}} \qquad \text{(Eq 5.14a)}$$

$$\sin\alpha = \frac{dp}{dl} = \phi'\sqrt{\frac{E}{H}} \qquad \text{(Eq 5.14b)}$$

$$\tan\alpha = \frac{dp}{ds} = \phi'\sqrt{\frac{E}{G}} \qquad \text{(Eq 5.14c)}$$

With these expressions and the relation $\tan\alpha = \sin\alpha/\cos\alpha$, Eq 5.12 can be transformed into:

$$\frac{\mathbf{C}'(t)}{\sqrt{\mathbf{C}'(t)\cdot\mathbf{C}'(t)}} = \frac{1}{\sqrt{H(t)}}\begin{Bmatrix} \rho'(t)\cos(\phi(t)) - \rho(t)\phi'(t)\sin(\phi(t)) \\ \rho'(t)\sin(\phi(t)) + \rho(t)\phi'(t)\cos(\phi(t)) \\ \zeta'(t) \end{Bmatrix}$$

$$= \begin{Bmatrix} -\left(\sqrt{\frac{E}{G}}\phi'(t)\right)\cos\alpha(t)\sin(\phi(t)) + \left(\frac{\rho'(t)}{\sqrt{G}}\right)\cos\alpha(t)\cos(\phi(t)) \\ +\left(\sqrt{\frac{E}{G}}\phi'(t)\right)\cos\alpha(t)\cos(\phi(t)) + \left(\frac{\rho'(t)}{\sqrt{G}}\right)\cos\alpha(t)\sin(\phi(t)) \\ \left(\frac{\zeta'(t)}{\sqrt{G}}\right)\cos\alpha(t) \end{Bmatrix}$$

$$= \begin{Bmatrix} -\sin\alpha(t)\sin(\phi(t)) + \left(\frac{\rho'(t)}{\sqrt{G}}\right)\cos\alpha(t)\cos(\phi(t)) \\ +\sin\alpha(t)\cos(\phi(t)) + \left(\frac{\rho'(t)}{\sqrt{G}}\right)\cos\alpha(t)\sin(\phi(t)) \\ \left(\frac{\zeta'(t)}{\sqrt{G}}\right)\cos\alpha(t) \end{Bmatrix} \qquad \text{(Eq 5.15)}$$

which is identical to Eq 5.10 for positive $\rho(t)$. The obtained expression can be further simplified by observing Fig. 5.3 once again. Following the direction of **t**, it can be seen that the height of the meridian profile $\zeta$ is increasing, but the radial distance $\rho$ is decreasing; hence, for $ds > 0$, $d\zeta > 0$ and $d\rho < 0$:

$$\cos\beta = \frac{d\zeta}{ds} = \frac{\zeta'(t)dt}{\sqrt{G}dt} = \frac{\zeta'(t)}{\sqrt{G}} \qquad \text{(Eq 5.16a)}$$

$$\sin\beta = -\frac{d\rho}{ds} = -\frac{\rho'(t)dt}{\sqrt{G}dt} = -\frac{\rho'(t)}{\sqrt{G}} \qquad \text{(Eq 5.16b)}$$

$$\tan\beta = -\frac{d\rho}{dz} = -\frac{\rho'(t)}{\zeta'(t)} \qquad \text{(Eq 5.16c)}$$

With the first two relations, the parenthesized terms in Eq 5.15 can be transformed into:

$$\frac{\mathbf{C}'(t)}{\sqrt{\mathbf{C}'(t)\cdot\mathbf{C}'(t)}} = \begin{Bmatrix} -\sin\alpha(t)\sin(\phi(t)) + \left(\frac{\rho'(t)}{\sqrt{G}}\right)\cos\alpha(t)\cos(\phi(t)) \\ +\sin\alpha(t)\cos(\phi(t)) + \left(\frac{\rho'(t)}{\sqrt{G}}\right)\cos\alpha(t)\sin(\phi(t)) \\ \left(\frac{\zeta'(t)}{\sqrt{G}}\right)\cos\alpha(t) \end{Bmatrix}$$

$$= \begin{Bmatrix} -\sin\alpha(t)\sin(\phi(t)) - \cos\alpha(t)\sin\beta(t)\cos(\phi(t)) \\ +\sin\alpha(t)\cos(\phi(t)) - \cos\alpha(t)\sin\beta(t)\sin(\phi(t)) \\ \cos\beta(t)\cos\alpha(t) \end{Bmatrix} \qquad \text{(Eq 5.17)}$$

This relation can immediately be verified by a simple geometric consideration (Fig. 5.2). At $\phi = 0$, the vector **T** is:

$$\mathbf{T}_{\phi=0} = \begin{Bmatrix} \frac{d\rho}{ds} \\ 0 \\ \frac{dz}{ds} \end{Bmatrix} = \begin{Bmatrix} -\sin\beta \\ 0 \\ \cos\beta \end{Bmatrix} \qquad \text{(Eq 5.18)}$$

Rotated over an angle $\phi$, the vector **T** will take the following form:

$$\mathbf{T} = \begin{bmatrix} \cos\phi & -\sin\phi & 0 \\ \sin\phi & \cos\phi & 0 \\ 0 & 0 & 1 \end{bmatrix} \cdot \begin{Bmatrix} -\sin\beta \\ 0 \\ \cos\beta \end{Bmatrix} = \begin{Bmatrix} -\sin\beta\cos\phi \\ -\sin\beta\sin\phi \\ \cos\beta \end{Bmatrix} \qquad \text{(Eq 5.19)}$$

At the same time, the **B** vector (Eq 5.5) is given by:

$$\mathbf{B} = \begin{Bmatrix} -\sin\phi \\ \cos\phi \\ 0 \end{Bmatrix} \qquad \text{(Eq 5.20)}$$

With Eq 5.19, it can be deduced that:

$$\mathbf{t} = \mathbf{T}\cos\alpha + \mathbf{B}\sin\alpha =$$

$$= \begin{Bmatrix} -\sin\beta\cos\phi \\ -\sin\beta\sin\phi \\ \cos\beta \end{Bmatrix}\cos\alpha + \begin{Bmatrix} -\sin\phi \\ \cos\phi \\ 0 \end{Bmatrix}\sin\alpha =$$

$$= \begin{Bmatrix} -\sin\alpha\sin\phi - \cos\alpha\sin\beta\cos\phi \\ +\sin\alpha\cos\phi - \cos\alpha\sin\beta\sin\phi \\ \cos\alpha\cos\beta \end{Bmatrix} \qquad \text{(Eq 5.21)}$$

which is identical to Eq 5.17.

Besides the verification of the obtained expressions, a significant advantage of this dual approach is the derivation of equations that relate the goniometric quantities $\{\alpha, \beta, \phi\}$ to the meridian parameterization $\{\rho(t), \zeta(t)\}$. The link between the meridian profile shape and the $\phi$ propagation of an overwound fiber is completely controlled by the winding angle $\alpha$. In this sense, the rather obvious trigonometric relations presented in Eq 5.14 are very important in filament winding.

To aid the derivations in the section associated with nongeodesic fiber paths, it is convenient to express the vectors, **T**, **B**, and **N**, in the same goniometric quantities $\{\alpha, \beta, \phi\}$ as **t**. Assuming $\rho(t) > 0$, substitution of Eq 5.16(a, b) into Eq 5.3, 5.5, and 5.8 leads to:

$$\mathbf{T} = \begin{Bmatrix} -\sin\beta\cos\phi \\ -\sin\beta\sin\phi \\ \cos\beta \end{Bmatrix}, \mathbf{B} = \begin{Bmatrix} -\sin\phi \\ \cos\phi \\ 0 \end{Bmatrix}, \mathbf{N} = \begin{Bmatrix} \cos\beta\cos\phi \\ \cos\beta\sin\phi \\ \sin\beta \end{Bmatrix} \qquad \text{(Eq 5.22)}$$

Note that while $\mathbf{C}'$ depends only on $t$ (Eq 5.17), the vector **N** depends on both $t$ and $\phi$ (normal vector of a surface). By assuming a function $\phi(t)$, **N** is entirely controlled by $t$ but does still represent the normal vector of the surface at a point belonging to the curve **C** (so it will not necessarily represent the normal vector of the curve itself). The vector, normal to a surface-supported curve, is generally not the same as the normal vector of that surface; see also the section on (non-) geodesic curves.

## The Role of the Winding Angle

For the theory presented here, one should realize that once the meridian profile is given, the angle $\beta$ becomes fully determined as a function of $t$ (Eq 5.16). To describe a winding path, a definition for $\phi = \phi(t)$ (and thereof $\phi'(t)$ or, alternatively $\alpha = \alpha(t)$, Eq 5.14c) is needed. Traditionally, a winding path is determined by $\alpha(t)$. For example, geodesic winding relies on (Ref 5.4, 5.5):

$$\sin(\alpha(t)) = \frac{c}{\rho(t)} \qquad \text{(Eq 5.23)}$$

where the minimum radius $c$ (usually referred to as the polar opening radius) is depicted in Fig. 5.2. For nongeodesic winding, the relation $\alpha(t)$ can only be derived by solving a nonlinear differential equation (next section). Assuming now that this function is known and using the definition of $G$ in Eq 5.4, the parallel angle $\phi(t)$ can be obtained by implementation of Eq 5.14(c):

$$\phi(t) = \int_{\tau=t_0}^{\tau=t} \frac{c}{\rho(\tau)}\sqrt{\frac{\rho'^2(\tau)+\zeta'^2(\tau)}{\rho^2(\tau)-c^2}}\,d\tau \qquad \text{(Eq 5.24)}$$

For geodesic winding, the aforementioned relation takes the following form:

$$\phi(t) = \int_{\tau=t_0}^{\tau=t} \frac{c}{\rho(\tau)}\sqrt{\frac{\rho'^2(\tau)+\zeta'^2(\tau)}{\rho^2(\tau)-c^2}}\,d\tau \qquad \text{(Eq 5.25)}$$

Note that for $\rho(t) = c$, the integrand becomes infinite, and $\alpha$ attains the value $\pi/2$ (Eq 5.23). This condition is required at the polar opening where the fiber bundle should be exactly tangent to the polar periphery to ensure continuation of the winding process.

From the aforementioned, it can be concluded that once the function $\alpha(t)$ is known, the fiber path becomes fully determined. This relation is simple for geodesic winding, but for nongeodesic winding, it is first necessary to consider the definitions of some curvatures.

## Main Shell Curvatures

As a curved path in three-dimensional (3-D) space is followed, the orientation (one's direction of walking over that curve) is continuously varying. For example, when proceeding along the meridian direction, the orientation of the tangent vector **T** will change (Fig. 5.4). This

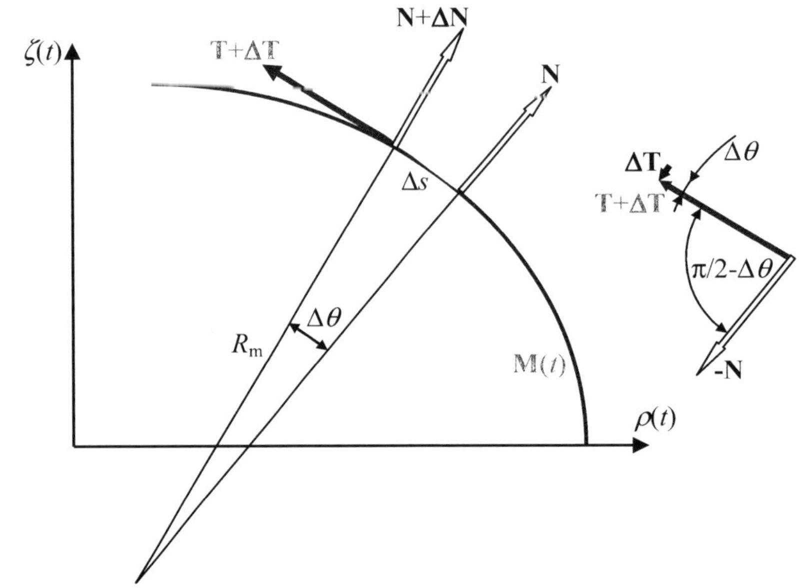


**Fig. 5.4** Intuitive consideration of the radius of curvature in meridional direction

modification in orientation can be expressed in an angle, say $\Delta\theta$, per unit of covered distance, $\Delta s$, along that path. The amount of steering is then $\Delta\theta/\Delta s$; this is referred to as the curvature, $k$. From the figure, it can be derived immediately that:

$$\Delta s = R_m \Delta\theta \rightarrow \frac{1}{R_m} = k_m = \frac{\Delta\theta}{\Delta s} \quad \text{(Eq 5.26)}$$

where the subscript "m" refers to the meridional direction.

A key observation is that the "new" orientation vector $\mathbf{T} + \Delta\mathbf{T}$ is not exactly vertical to $\mathbf{N}$ anymore (Fig. 5.4). In essence, this nonverticality provides the degree of steering. According to the definition of a vector inner product (note that both participating vectors have a unit length):

$$\begin{aligned}\cos\left(\frac{\pi}{2} - \Delta\theta\right) &= \sin\Delta\theta = -\mathbf{N}\cdot(\mathbf{T}+\Delta\mathbf{T}) \\ &= -\mathbf{N}\cdot\mathbf{T} - \mathbf{N}\cdot\Delta\mathbf{T} = 0 - \mathbf{N}\cdot\Delta\mathbf{T} \\ &\rightarrow \sin\Delta\theta \approx \Delta\theta = -\mathbf{N}\cdot\Delta\mathbf{T} = -\mathbf{N}\cdot\mathbf{T}_t\Delta t\end{aligned} \quad \text{(Eq 5.27)}$$

With Eq 5.26, one obtains:

$$k_m = \frac{d\theta}{ds} = \frac{-\mathbf{N}\cdot\mathbf{T}_t dt}{\sqrt{G}dt} = -\frac{\mathbf{N}\cdot\mathbf{T}_t}{\sqrt{G}} \quad \text{(Eq 5.28)}$$

where Eq 5.3, the definition of $G$, has been substituted.

At, for example, $\phi = 0$, the vector $\mathbf{N}$ will be of the form $\mathbf{N} = \{x, y, z\} = \{+, 0, +\}$. On the contrary, the vector $\mathbf{T}$ is (Fig. 5.4) $\Delta\mathbf{T} = \{-, 0, -\}$; hence, $k_m$ will become positive. Traditionally, in 3-D geometry, positive curvatures are associated with convex bodies. Concave bodies (hollow shapes) correspond to negative curvatures. This practice is opposite to two-dimensional (2-D) geometry of curves where, for example, a $\zeta(\rho)$ curve corresponding to the convex meridian example would have a negative second derivative. The curvature, in this case given by the classic equation $k_m = \zeta''(\rho)/(1 + \zeta'^2(\rho))^{3/2}$, is negative (Ref 5.6).

An additional reason for demanding positive curvatures for convex 3-D bodies can be formulated as follows. Examination of Fig. 5.4 reveals that in the direction of $\mathbf{T}$ the angle $\Delta\theta$ is increasing (in the mathematical positive direction, counterclockwise); hence, $k_m$ should be positive (for an outward-pointing normal vector $\mathbf{N}$). Nevertheless, classical textbooks on differential geometry employ the 2-D-inspired definition of curvature without the minus sign, as presented in Eq 5.28.

A presentation of $k_m$ in vector form looks elegant, but it still must be related to the original shape parameterization. Substitution of Eq 5.3 and 5.8 into Eq 5.28 leads to:

$$k_m(t) = \frac{\rho'(t)\zeta''(t) - \rho''(t)\zeta'(t)}{\underbrace{\left(\rho'^2(t) + \zeta'^2(t)\right)^{3/2}}_{G^{3/2}}} \quad \text{(Eq 5.29)}$$

where ρ(*t*) is assumed positive.

In differential geometry, the product $\mathbf{N} \cdot \mathbf{T}_t$, denoted by *g* or *N*, is one of the three coefficients of the so-called second fundamental form (Ref 5.2–5.6):

$$e, L = \mathbf{S}_{\phi\phi} \cdot \mathbf{N} = \mathbf{B}_\phi \cdot \mathbf{N}\sqrt{E} = \mathbf{S}_{\phi\phi} \cdot \frac{\mathbf{B} \times \mathbf{T}}{\|\mathbf{B} \times \mathbf{T}\|} = \frac{\det\left[\mathbf{S}_{\phi\phi}, \mathbf{S}_\phi, \mathbf{S}_t\right]}{\sqrt{EG}} \quad \text{(Eq 5.30a)}$$

$$f, M = \mathbf{S}_{t\phi} \cdot \mathbf{N} = 0 \quad \text{(Eq 5.30b)}$$

$$g, N = \mathbf{S}_{tt} \cdot \mathbf{N} = \mathbf{T}_t \cdot \mathbf{N}\sqrt{G} = \mathbf{S}_{tt} \cdot \frac{\mathbf{B} \times \mathbf{T}}{\|\mathbf{B} \times \mathbf{T}\|} = \frac{\det\left[\mathbf{S}_{tt}, \mathbf{S}_\phi, \mathbf{S}_t\right]}{\sqrt{EG}} \quad \text{(Eq 5.30c)}$$

where the provided expressions apply only on shells of revolution. The coefficients of the first fundamental form provide information about metrics on a surface, while the second fundamental-form parameters reveal how the shape orientation changes with {*t*, ϕ}. In fact, the normal vector **N** can be regarded as a shape operator (Ref 5.2, 5.3). In classical differential geometry textbooks, the curvature in meridional direction, $k_m$, is defined as:

$$k_m = \frac{g}{G} \quad \text{(Eq 5.31)}$$

which coincides exactly with Eq 5.28, after multiplication with –1. By following the same reasoning for the parallel direction (ϕ varies, while *t* is fixed), one can write:

$$k_p = \frac{e}{E} \quad \text{(Eq 5.32)}$$

The parallel curvature (= $1/R_p$) is associated with the directional change of the vector **B** (Fig. 5.5). Once again, this change in direction is

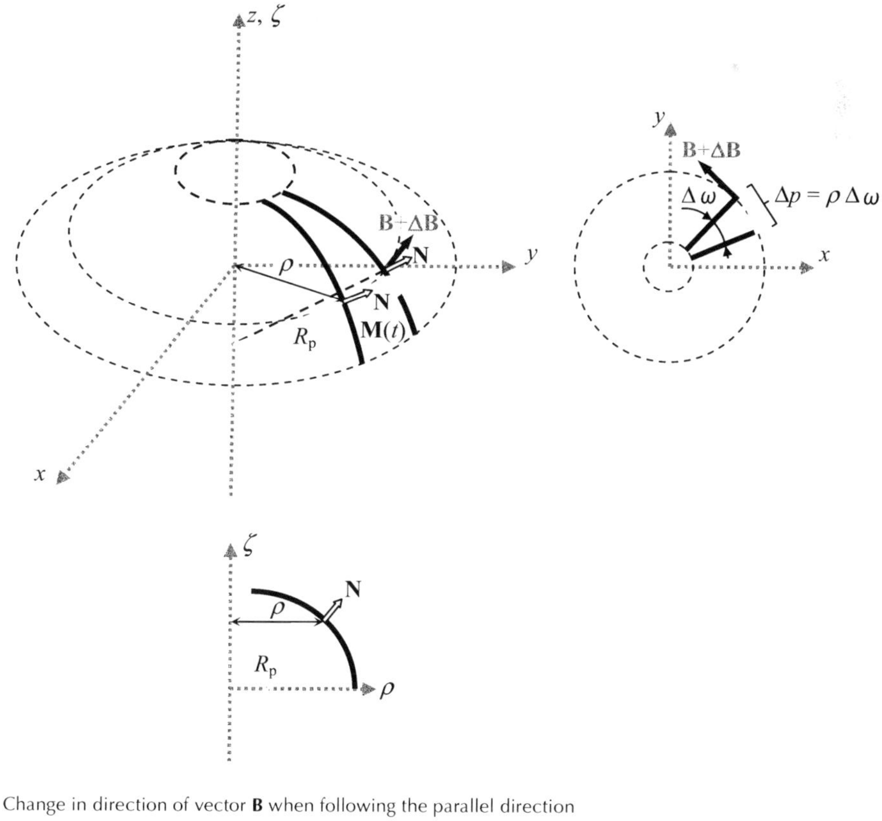


**Fig. 5.5** Change in direction of vector **B** when following the parallel direction

measured with reference to the normal vector **N**. The orientation change, $\Delta\omega$, takes place over a peripheral distance $\Delta p = R_p\Delta\omega$, where $1/R_p = k_p$ is the parallel curvature. The derivation of $k_p$ is similar to that of $k_m$, with $\mathbf{T}_t$ replaced by $\mathbf{B}_\phi$ and $ds$ by $dp$:

$$k_p = \frac{d\omega}{dp} = \frac{-\mathbf{N}\cdot\mathbf{B}_\phi d\phi}{\sqrt{E}d\phi} = \frac{-\mathbf{N}\cdot\mathbf{B}_\phi}{\sqrt{E}} \qquad \text{(Eq 5.33)}$$

which is in accordance with Eq 5.30(a) when multiplied by –1. For the term $dp$, Eq 5.6 was used. Substitution of Eq 5.5 and 5.8 in the $k_p$ expression finally leads to:

$$k_p(t) = \frac{\zeta'(t)}{|\rho(t)|\sqrt{\rho'^2(t)+\zeta'^2(t)}} = \frac{\zeta'(t)}{\sqrt{GE}}, \quad (\rho(t)>0) \qquad \text{(Eq 5.34)}$$

With Eq 5.4 and 5.16(a), it is easy to deduce that ($\rho(t) > 0$):

$$k_p(t)\sqrt{E} = \frac{\zeta'(t)}{\sqrt{G}}\frac{dt}{dt} = \frac{d\zeta}{ds} = \cos\beta \qquad \text{(Eq 5.35)}$$

This is why the $R_p$-axis in Fig. 5.5 ends at exactly its intersection with the $\zeta$-axis. Apparently, for a convex shell of revolution, the axis of rotational symmetry is also an axis of focal points.

## Curvature of a Fiber Trajectory on a Shell of Revolution

As previously observed, the two main curvatures, $k_m$ and $k_p$, are related to an angular deviation over a distance in, respectively, the meridional and parallel direction (Fig. 5.6):

$$k_m = \frac{d\theta}{ds}$$
$$k_p = \frac{d\omega}{dp} \qquad \text{(Eq 5.36)}$$

Similarly, the curvature (normal to the shell surface) of the fiber path is given by (Fig. 5.6):

$$k_n = \frac{d\psi}{dl} \qquad \text{(Eq 5.37)}$$

where $dl$ is the length increment in the fiber direction.

As implemented for the tangent vector **t** (Eq 5.21), the fiber-related angular increment $\Delta\psi$ over $dl$ can also be related to $d\theta$ and $d\omega$:

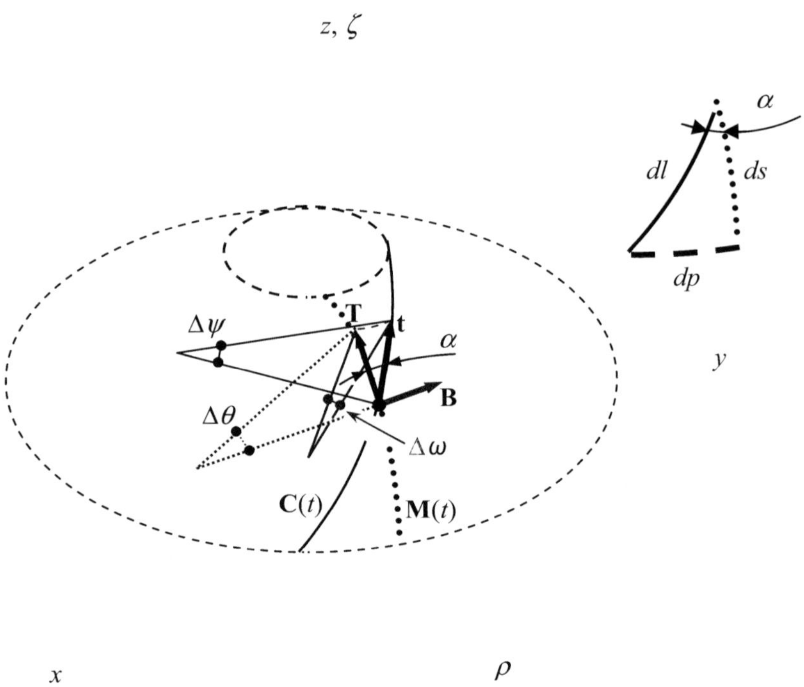


**Fig. 5.6** Curvature of curve **C** as related to the two main curvatures of the shell

$$k_n = \frac{d\psi}{dl} = \frac{\cos\alpha d\theta + \sin\alpha d\omega}{dl} \quad \text{(Eq 5.38)}$$

With Eq 5.14 and the expressions for $k_m$ and $k_p$, one can immediately conclude that:

$$k_n = \frac{d\psi}{dl} = \frac{d\theta}{\left(\frac{ds}{\cos\alpha}\right)}\cos\alpha + \frac{d\omega}{\left(\frac{dp}{\sin\alpha}\right)}\sin\alpha$$
$$= k_m \cos^2\alpha + k_p \sin^2\alpha \quad \text{(Eq 5.39)}$$

This is the well-known Euler curvature relation (Ref 5.1–5.3). It must be emphasized again that the obtained result represents the curvature of the fiber path in the direction normal to the surface (thus in the direction of **N**).

Next to this component, the fiber path may show some in-plane steering; this is captured by the so-called geodesic curvature parameter. The actual curvature (referred to as total curvature, $k_t$) is obtained by a quadratic summation of $k_n$ and $k_g$ or, more formally, by the implementation of $\mathbf{C}''(t) \cdot \mathbf{b}$, where $\mathbf{b} = \mathbf{t} \times \mathbf{N}$ (see Eq 5.11 for the fiber path parameterization $\mathbf{C}(t)$). In essence, $\mathbf{C}''(t)$ is the vector normal to the fiber path and not necessarily to the surface; it defines, at least locally, the plane in which the path is located. For $k_g = 0$, the normal vector of the path itself, $\mathbf{C}''(t)$, coincides with **N** (normal of the surface), so there is no in-plane steering. A detailed derivation is provided in the next section.

## Reference Frame at a Curve

Earlier, an important distinction was made between the normal vector of a curve, $\mathbf{C}''$, and the normal vector on a surface, **N**. For convenience, the definitions of a shell of revolution and a curve on it are repeated here:

$$\mathbf{S}(t,\phi) = \begin{Bmatrix} \rho(t)\cos\phi \\ \rho(t)\sin\phi \\ \zeta(t) \end{Bmatrix}, \quad \begin{matrix} t_0 \le t \le t_1 \\ 0 \le \phi \le 2\pi \end{matrix} \quad \text{(Eq 5.40a)}$$

$$\mathbf{C}(t) = \begin{Bmatrix} \rho(t)\cos(\phi(t)) \\ \rho(t)\sin(\phi(t)) \\ \zeta(t) \end{Bmatrix}, \quad t_0 \le t \le t_1 \quad \text{(Eq 5.40b)}$$

The normal vector **N** of the surface is a function of $\{t, \phi\}$ (Eq 5.22). For a curve placed on that surface where $\phi = \phi(t)$, the normal vector **N** becomes:

$$\mathbf{N}(t) = \frac{1}{\sqrt{G}} \begin{Bmatrix} \zeta'(t)\cos\phi(t) \\ \zeta'(t)\sin\phi(t) \\ -\rho'(t) \end{Bmatrix} = \begin{Bmatrix} \cos\beta(t)\cos\phi(t) \\ \cos\beta(t)\sin\phi(t) \\ \sin\beta(t) \end{Bmatrix}$$
(Eq 5.41)

Although it depends only on $t$, it does still represent the normal vector of the surface **S** (and not the normal vector of the curve **C**) for a surface point that belongs to that curve.

The vector normal to the curve defines locally a plane in which the curve lies (gray plane in Fig. 5.7). This normal is given by differentiation of the tangent vector $\mathbf{t}(t)$. For convenience, Eq 5.15 and 5.21 are repeated:

$$\mathbf{t}(t) = \frac{\mathbf{C}'(t)}{\sqrt{\mathbf{C}'(t)\cdot\mathbf{C}'(t)}} =$$
$$= \frac{1}{\underbrace{\sqrt{G(t)+E(t)\phi'^2(t)}}_{\sqrt{H}}} \begin{Bmatrix} \rho'(t)\cos(\phi(t)) - \rho(t)\phi'(t)\sin(\phi(t)) \\ \rho'(t)\sin(\phi(t)) + \rho(t)\phi'(t)\cos(\phi(t)) \\ \zeta'(t) \end{Bmatrix}$$

$$= \begin{Bmatrix} -\sin\alpha(t)\sin\phi(t) - \cos\alpha(t)\sin\beta(t)\cos\phi(t) \\ +\sin\alpha(t)\cos\phi(t) - \cos\alpha(t)\sin\beta(t)\sin\phi(t) \\ \cos\alpha(t)\cos\beta(t) \end{Bmatrix}$$
(Eq 5.42)

Next to the tangent vector **t**, a vector **b** is depicted (Fig. 5.7), located in the surface plane and oriented vertical to **t**. The **b** vector is given by:

$$\mathbf{b} = \mathbf{t}\times\mathbf{N} = -\mathbf{T}\sin\alpha + \mathbf{B}\cos\alpha$$

$$= \begin{Bmatrix} \sin\beta\cos\phi \\ \sin\beta\sin\phi \\ -\cos\beta \end{Bmatrix}\sin\alpha + \begin{Bmatrix} -\sin\phi \\ \cos\phi \\ 0 \end{Bmatrix}\cos\alpha$$

$$= \begin{Bmatrix} -\cos\alpha\sin\phi + \sin\alpha\sin\beta\cos\phi \\ +\cos\alpha\cos\phi + \sin\alpha\sin\beta\sin\phi \\ -\sin\alpha\cos\beta \end{Bmatrix} \quad \text{(Eq 5.43)}$$

where **T** and **B** are given in Eq 5.22. At this point, all important vectors and differentials are derived.

The key mechanism for quantifying the geodesic curvature is as follows. If the fiber is indeed

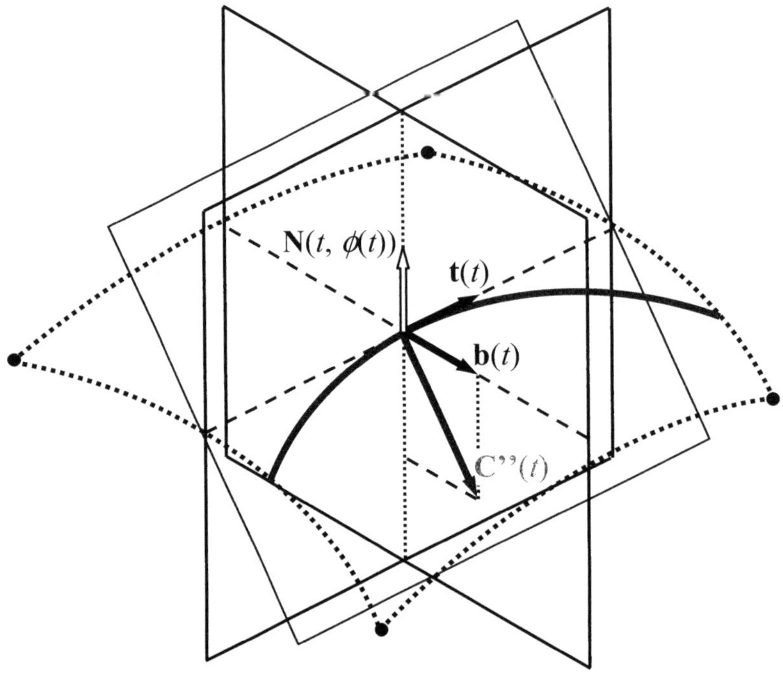


**Fig. 5.7** The vector **C″** normal to a fiber path does not necessarily coincide with the surface normal **N** at that point.

following the shortest possible path between two points, it will not show any in-plane steering (like driving a car on the globe from *A* to *B* without cornering). The **C″** vector, describing the change in the orientation of **C′** and therefore the amount of steering, will, in this case, have only a component vertical to the surface (road); in other words, **N** and **C″** are exactly parallel, and **N** · **C″** = 1. In every other case, **C″** will have a component in the direction of **b** (in the road plane), expressed by the condition **b** · **C″** ≠ 0. Consequently, there will be some in-plane steering whose magnitude is given by the geodesic curvature, $k_g$. Before attempting to quantify this, it is convenient to first revisit the definition of the normal curvature.

## Normal Curvature of a Curve Revisited

The normal curvature of a path ($k_n$) was derived in the previous section. It is considered here in a more formal way, as a part of the efforts to finally come up with an expression for $k_g$. Similar to the derivations for $k_m$ (Eq 5.28) and $k_p$ (Eq 5.33), the expression for $k_n$ (Fig. 5.7, 5.8) can be written as:

$$k_n = \frac{\Delta\psi}{\Delta l} = -\frac{\mathbf{N}\cdot\mathbf{C}''}{\sqrt{H}} \qquad \text{(Eq 5.44)}$$

where **N** · **C″** expresses the directional change of **C′** with respect to **N**; this corresponds to an angle $d\psi$ over a distance $dl$ (Fig. 5.6, Eq 5.37). Note that $dl = H^{1/2}dt$ (Eq 5.13). With **C″** = **t′** and Eq 5.41 and 5.42, the expression for $k_n$ becomes:

$$k_n = -\frac{\mathbf{N}\cdot\mathbf{C}''}{\sqrt{H}} = -\frac{\mathbf{N}\cdot\mathbf{t}'}{\sqrt{H}}$$

$$= -\frac{1}{\sqrt{H}}\begin{Bmatrix}\cos\beta\cos\phi\\ \cos\beta\sin\phi\\ \sin\beta\end{Bmatrix} \cdot \frac{d}{dt}\left(\begin{Bmatrix}-\sin\alpha\sin\phi-\cos\alpha\sin\beta\cos\phi\\ +\sin\alpha\cos\phi-\cos\alpha\sin\beta\sin\phi\\ \cos\alpha\cos\beta\end{Bmatrix}\right)$$

$$= -\frac{1}{\sqrt{H}}(-\beta'\cos\alpha-\phi'\sin\alpha\cos\beta) \qquad \text{(Eq 5.45)}$$

The difficulty now is to eliminate the angular parameters. Using the definition of $G$ (Eq 5.4) and the expression for $k_m$ (Eq 5.29), Eq 5.16(c) becomes:

$$\tan\beta = -\frac{\rho'(t)}{\zeta'(t)} \to \beta = \arctan\left(-\frac{\rho'}{\zeta'}\right)$$

$$\to \beta' = \frac{\rho'\zeta'' - \rho''\zeta'}{\rho'^2 + \zeta'^2} = k_m\sqrt{G} \qquad \text{(Eq 5.46)}$$

In addition, the term $\cos\beta$ in Eq 5.45 can be replaced by $k_p E^{1/2}$ (Eq 5.35):

$$k_n = \frac{1}{\sqrt{H}}(\beta'\cos\alpha + \phi'\sin\alpha\cos\beta)$$

$$= k_m\sqrt{\frac{G}{H}}\cos\alpha + k_p\phi'\sin\alpha\sqrt{\frac{E}{H}} \qquad \text{(Eq 5.47)}$$

With Eq 5.14(a, b), the following is derived:

$$k_n = k_m\cos\alpha\cos\alpha + k_p\sin\alpha\sqrt{\frac{H}{E}}\sin\alpha\sqrt{\frac{E}{H}} =$$

$$= k_m\cos^2\alpha + k_p\sin^2\alpha$$

$$\text{(Eq 5.48)}$$

which is identical to Eq 5.39. The normal curvature is one of the two components needed to implement nongeodesic trajectories.

## Geodesic Curvature

In contrast to the elaborated derivations found in the literature, the determination of the geodesic curvature $k_g$ is easier than that of $k_n$. With a motivation similar to that of $k_n$ and Fig. 5.7 and 5.8, the geodesic curvature $k_g$ can be expressed as follows:

$$k_g = \frac{\Delta\chi}{\Delta l} = \frac{\mathbf{b}\cdot\mathbf{C}''}{\sqrt{H}} = \frac{1}{\sqrt{H}}(\alpha' - \phi'\sin\beta) \qquad \text{(Eq 5.49)}$$

Note that, contrary to the $k_n$ derivation, the reference vector **b** is pointing to the center of curvature; therefore, the minus sign is neglected. With Eq 5.14(c) and Eq 5.16(b), the expression for $k_g$ becomes:

$$k_g = \frac{1}{\sqrt{H}}\left(\alpha' + \frac{\rho'}{\sqrt{E}}\tan\alpha\right) \qquad \text{(Eq 5.50)}$$

The derivative $\rho'$ can be related to $E$ with the aid of Eq 5.16, ($\rho(t) > 0$):

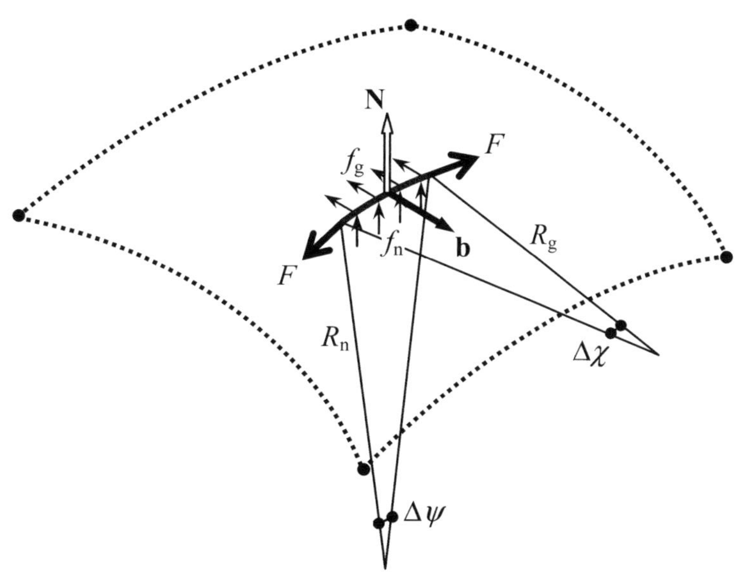


**Fig. 5.8** Equilibrium of forces at an elementary piece of a tensioned fiber path, placed on a convex surface

$$\rho(t)=\sqrt{E(t)} \rightarrow \rho'(t)=\frac{1}{2}\frac{E'(t)}{\sqrt{E(t)}} \quad \text{(Eq 5.51)}$$

Plugging this relation into Eq 5.50 leads to:

$$k_g = \frac{1}{\sqrt{H}}\left(\alpha' + \frac{1}{2}\frac{E'}{E}\tan\alpha\right) \quad \text{(Eq 5.52)}$$

With Eq 5.14(a), this relation can also be written as:

$$k_g = \frac{\cos\alpha}{\sqrt{G}}\left(\alpha' + \frac{1}{2}\frac{E'}{E}\frac{\sin\alpha}{\cos\alpha}\right) = \alpha'\frac{\cos\alpha}{\sqrt{G}} + \frac{1}{2}\frac{E'}{E}\frac{\sin\alpha}{\sqrt{G}} \quad \text{(Eq 5.53)}$$

This is the most popular form, and it is usually referred to as the Liouville equation (Ref 5.1–5.3). As previously demonstrated, a geodesic path is characterized by $k_g = 0$. This implies that (Eq 5.52):

$$k_g = \alpha' + \left(\frac{E'}{2E}\right)\frac{\sin\alpha}{\cos\alpha}$$

$$= 0 \rightarrow \alpha'\cos\alpha + \left(\frac{E'}{2E}\right)\sin\alpha = 0$$

$$\rightarrow (\sin\alpha)' + \frac{1}{2}\left(\frac{E'}{E}\right)\sin(\alpha)$$

$$= 0 \rightarrow \frac{(\sin\alpha)'}{\sin\alpha} = -\frac{1}{2}\left(\frac{E'}{E}\right)$$

$$\rightarrow \left[\ln(\sin\alpha)\right]' = -\left[\ln\left(\sqrt{E}\right)\right]' \rightarrow \ln(\sin\alpha)$$

$$= -\ln\left(\sqrt{E}\right) + \ln(\text{const.})$$

$$\rightarrow \sin\alpha = \frac{\text{const.}}{\sqrt{E}} = \frac{\text{const.}}{\rho} \quad \text{(Eq 5.54)}$$

which is the well-known Clairaut equation (Ref 5.1–5.3). Because the winding angle at the polar opening ($\rho = c$) is equal to $\pi/2$ (Fig. 5.2), the constant typically represents the polar opening radius, $c$.

## Winding with Nongeodesics

In the previous section, it was deduced that a geodesic, the shortest line connecting two points on a surface, does not show any in-plane steering. Driving from $A$ to $B$ without cornering does certainly provide the shortest route. Any deviation from this path will require a lateral force; cornering of a car relies on the lateral friction between road and front tires.

Because the (tensioned) fiber is placed on a convex surface (Fig. 5.8), the normal force per unit length can be derived by considering the equilibrium of forces in the direction of $\mathbf{N}$:

$$f_n R_n \Delta\psi = 2F\sin\left(\frac{\Delta\psi}{2}\right) \quad F\Delta\psi \rightarrow f_n = \frac{F}{R_n}$$

(Eq 5.55)

where $R_n = 1/k_n$. The equilibrium of forces in the direction of $\mathbf{b}$ leads to:

$$f_g R_g \Delta\chi = 2F\sin\left(\frac{\Delta\chi}{2}\right) \quad F\Delta\chi \rightarrow f_g = \frac{F}{R_g} \quad \text{(Eq 5.56)}$$

with $R_g = 1/k_g$. To avoid fiber sliding, the lateral force $f_g$ should not exceed the limit defined by the available static friction, $\mu_{max}$ (for the fiber ↔ surface interface):

$$|f_g| \le |\mu_{max}||f|_n \rightarrow \left|\frac{f_g}{f_n}\right| \le |\mu_{max}| \quad \text{(Eq 5.57)}$$

Or, in terms of curvatures:

$$\left|\frac{k_g}{k_n}\right| \le |\mu_{max}| \quad \text{(Eq 5.58)}$$

To consider the most generic case, friction is introduced as a function of $\mu(t)$, where $|\max[\mu(t)]| \le |\mu_{max}|$. Hence, for a given meridian shape, the permissible in-plane steering ($k_g$) also becomes a function of $t$. With Eq 5.58, 5.48, and 5.14(a), the permissible in-plane steering becomes:

$$k_g = \mu k_n \rightarrow$$

$$\rightarrow \alpha' + \frac{1}{2}\frac{E'}{E}\tan\alpha = \mu k_n\sqrt{H} \rightarrow$$

$$\rightarrow \alpha' = -\frac{1}{2}\frac{E'}{E}\tan\alpha + \mu k_n\sqrt{H}$$

$$= -\frac{1}{2}\frac{E'}{E}\frac{\sin\alpha}{\cos\alpha} + \mu k_n\frac{\sqrt{G}}{\cos\alpha} \rightarrow$$

$$\rightarrow (\sin\alpha)' = -\frac{1}{2}\frac{E'}{E}\sin\alpha + \mu k_n\sqrt{G} \rightarrow$$

$$\rightarrow \frac{(\sin\alpha)'}{\sin\alpha} = -\frac{1}{2}\frac{E'}{E} + \frac{\mu k_n\sqrt{G}}{\sin\alpha} \quad \text{(Eq 5.59)}$$

Let the friction distribution be given by:

$$\mu(t) = m(t)\sin\alpha(t) \quad \text{(Eq 5.60)}$$

With this information, Eq 5.59 can be transformed into:

$$\frac{(\sin\alpha)'}{\sin\alpha} = -\frac{1}{2}\frac{E'}{E} + mk_n\sqrt{G} \rightarrow \left[\ln(\sin\alpha)\right]' = -\left[\ln\left(\sqrt{E}\right)\right]' + mk_n\sqrt{G}$$

$$\ln(\sin\alpha) = -\ln\left(\sqrt{E}\right) + \underbrace{\int_{\tau=0}^{t} mk_n\sqrt{G}\,d\tau}_{J(t)} + \ln c$$

$$\rightarrow \sin\alpha = \frac{ce^{J(t)}}{\sqrt{E}} \qquad \text{(Eq 5.61)}$$

For $m(t) = 0$, the Clairaut solution ($e^0 = 1$) is acquired. However, this equation is restricted to convex surfaces (Fig. 5.8); therefore, $k_n$ is always nonnegative. For flat surfaces, $k_n = 0$, the friction does not have any influence on the winding angle (path orientation); it is impossible to alternate the orientation of the (straight) fiber placed on a flat surface by means of friction, because there are no normal forces. For every positive $m(t)$ (and thereof positive $\mu(t)$), the winding angle will increase because $G^{1/2}$ is always positive. For negative friction values, the winding angle will become relatively smaller as compared to the geodesic one.

To maximize the effect of friction, it is usually assumed that $\mu = \mu_{max}$. In this case, it is generally difficult to set up an analytical solution (with an exception for conical surfaces) unless the friction function is formalized in such a way that Eq 5.59 has a straightforward solution. However, because the expression for $H$ may become complicated, it is usually more convenient to numerically solve the Liouville equation (Eq 5.53) in combination with Eq 5.58:

$$k_g = \alpha'\frac{\cos\alpha}{\sqrt{G}} + \frac{1}{2}\frac{E'}{E}\frac{\sin\alpha}{\sqrt{G}} = \mu k_n$$

$$\rightarrow \alpha' = \frac{\sqrt{G}}{\cos\alpha}\mu\left(k_m\cos^2\alpha + k_p\sin^2\alpha\right) - \frac{1}{2}\frac{E'}{E}\frac{\sin\alpha}{\sqrt{G}}\frac{\sqrt{G}}{\cos\alpha}$$

$$\rightarrow \alpha'\cos\alpha = \mu\sqrt{G}\left[k_m\left(1-\sin^2\alpha\right) + k_p\sin^2\alpha\right] - \frac{1}{2}\frac{E'}{E}\sin\alpha \rightarrow$$

$$\rightarrow (\sin\alpha)' = \left[\mu\sqrt{G}\left(k_p - k_m\right)\right]\sin^2\alpha + \left[-\frac{1}{2}\frac{E'}{E}\right]\sin\alpha + \left[\mu\sqrt{G}k_m\right]$$

$$\rightarrow f'(t) = \left[c_2(t)\right]f^2(t) + \left[c_1(t)\right]f(t) + \left[c_0(t)\right] \qquad \text{(Eq 5.62)}$$

With an initial value $\alpha(t_0)$, Eq 5.62 can be solved numerically by, for example, the fourth-order Runge-Kutta method over an interval $[t_0, t_1]$ (Ref 5.7, 5.8). However, care must be taken not to run into $\alpha = 0$ or $\alpha = \pi/2$ before reaching $t_1$. In this case, the equation becomes either singular (infinite $\alpha'$) or static ($\alpha' = 0$). The same warning applies for a solution based on Eq 5.61 (Ref 5.4).

## Example

One of the few analytical solutions available reflects on nongeodesic trajectories on conical shells of revolution. Because conical rings are widely used for the discretization of shells, these solutions are of great importance in engineering practice. Let a conical shell be parameterized as follows (Fig. 5.9):

$$\mathbf{M}(t) = \begin{Bmatrix}\rho(t)\\ \zeta(t)\end{Bmatrix} = S\begin{Bmatrix}u-(\sin\gamma)t\\ v+(\cos\gamma)t\end{Bmatrix},\quad \mathbf{S}(t,\phi) = s\begin{Bmatrix}\rho(t)\cos\phi\\ \rho(t)\sin\phi\\ \zeta(t)\end{Bmatrix}$$

$$0 \le t \le 1, \qquad 0 \le \phi \le 2\pi \qquad \text{(Eq 5.63)}$$

where $S$ is a scaling parameter with the dimension of length. For $\gamma = \pi/2$, the shell becomes an annulus; for $\gamma = 0$, a cylinder is obtained. Substitution of the previously given parameterization into Eq 5.59 results in:

$$\frac{(\sin\alpha)'}{\sin\alpha} = \frac{\sin\gamma + \mu\cos\gamma\sin\alpha}{u - t\sin\gamma} \qquad \text{(Eq 5.64)}$$

At the smallest radius (polar opening), the winding angle must be equal to $\pi/2$: $\alpha(1) = \pi/2$. With this condition, the solution of the differential equation becomes:

$$\sin\alpha\big|_{\alpha(1)=\pi/2} = \frac{u-\sin\gamma}{u-(\sin\gamma)t+\mu(\cos\gamma)(1-t)} = \frac{\rho(1)}{\rho(t)+\mu\left[\zeta(1)-\zeta(t)\right]} \qquad \text{(Eq 5.65)}$$

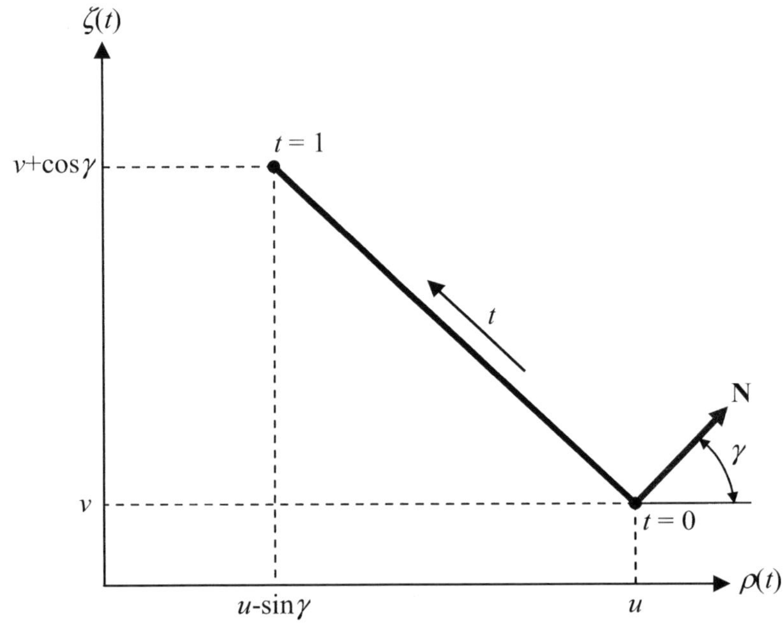


**Fig. 5.9** Parametric definition of a meridian for a conical shell of revolution

where it is clear that for $\mu = 0$, the solution will reduce to the Clairaut equation. Note that, depending on the sign of $\mu$, the winding angle can increase or decrease, amplified by $[\zeta(1) - \zeta(t)]$ (Fig. 5.10). When this height difference is zero (annulus), the friction does not have any influence on the resulting fiber path. This observation is completely in line with the comments after Eq 5.61. For a cylinder, $\gamma = 0$, the winding angle becomes equal to the boundary value $\alpha(1) = \pi/2$ and remains constant. (For the boundary value implemented here, the $\phi(t)$ function becomes infinite.) Once the winding angle at the boundary is chosen (and there is no friction), it will remain constant. In a more general consideration, an important restriction for all conical shells is that the function $\alpha(t)$ must remain in the interval $[0, \pi/2]$ for $t \in [0, 1]$. Therefore, the friction coefficient must comply with:

$$\sin\alpha(0) \le 1 \rightarrow \frac{u - \sin\gamma}{u + \mu\cos\gamma} \le 1 \rightarrow \mu_{\text{admissible}} \ge -\tan\gamma$$

(Eq 5.66)

This condition states that during the placement of a fiber on an inclined surface, for example, winding of hoop layers on a conical shell, the available friction must be greater than the tangent of that surface.

Entirely governed by $\alpha(t)$, the angle $\phi(t)$ will show a particular development as a function of $t$. According to the meridian parameterization (Eq 5.40b), this function is able to completely determine the 3-D geometry of a (geodesic or not) fiber path. To calculate this function, Eq 5.24 must be further examined:

$$\phi(t) = \int_{\tau=t_0}^{\tau=t} \sqrt{\frac{G(t)}{E(t)}} \tan(\alpha(\tau))\,d\tau = \frac{2}{\sqrt{mp}} \arctan\left(\sqrt{\frac{m(1-t)}{2u - m - pt}}\right)$$

$$\text{where} \begin{pmatrix} m = \sin\gamma - \mu\cos\gamma \\ p = \sin\gamma + \mu\cos\gamma \end{pmatrix} \text{and} \left(\alpha(1) = \frac{\pi}{2}\right)$$

(Eq 5.67)

where proper care must be taken to avoid extracting the square root of a negative argument. For cylinders and annuli, the $\phi(t)$ function reduces to:

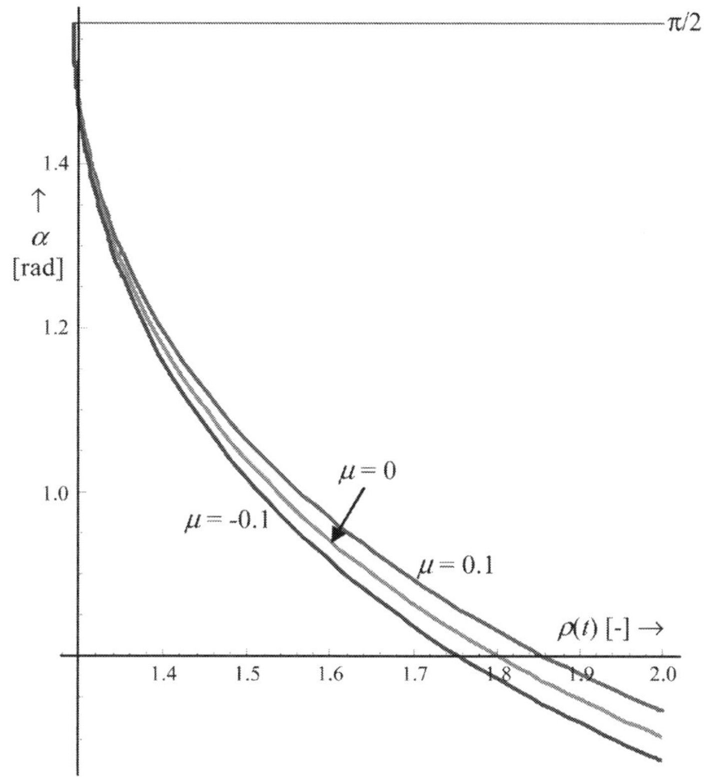


**Fig. 5.10** Influence of the friction on fiber paths, placed on a conical shell with $\gamma = \pi/4$

$\left(\alpha(1) = \frac{\pi}{2}, \gamma = 0, \mu \neq 0\right)$:

$$\phi(t) = \frac{2}{\mu}\arctan\left(\sqrt{\frac{\mu(1-t)}{2u+\mu(1-t)}}\right)$$

$\left(\alpha(1) = \frac{\pi}{2}, \gamma = \frac{\pi}{2}\right)$:

$$\phi(t) = 2\arctan\left(\sqrt{\frac{1-t}{2u-1-t}}\right) \quad \text{(Eq 5.68)}$$

Some results are provided in Fig. 5.11. On a cylinder with $\mu = 0$ and a polar winding angle $\alpha(1) = \pi/2$, the fiber will just describe a horizontal ring, infinite times. For a different initial winding angle, $\alpha_{\text{init}}$, the $\phi(t)$ function becomes:

$$\left(\gamma = 0, \mu = 0, \alpha_{\text{init}} \neq \frac{\pi}{2}\right): \ \phi(t) = (\tan\alpha_{\text{init}})\frac{t}{u} \quad \text{(Eq 5.69)}$$

which represents a helix.

Note that the other analytical $\alpha(t)$ and $\phi(t)$ solutions presented are based on the condition $\alpha(1) = \pi/2$. To discretize a given shell into a series of conical segments with a continuous winding angle development from one segment to the other, one must be able to solve Eq 5.64 for every possible initial condition $\alpha(T) = \alpha_{\text{init}}$ for $T = 1$ or $T = 0$ (respectively, for a prescribed winding angle at the top or bottom periphery of the conical segment, Fig. 5.9). In a more general fashion, the winding angle may also be prescribed somewhere halfway on the conical segment. For $0 \leq T \leq 1$, the solution of Eq 5.64 becomes:

$$\sin\alpha(t) = \frac{\rho(T)\sin\alpha_{\text{init}}}{\rho(t) + \mu\sin\alpha_{\text{init}}\left[\zeta(T) - \zeta(t)\right]} \quad \text{(Eq 5.70)}$$

With $\{\rho(T), \zeta(T)\} = \{SR, SZ\}$, $\sin\alpha(T) = \sin\alpha_{\text{init}} = Sc/\rho(T) = (Sc)/(SR)$, $k = \mu\sin\alpha_{\text{init}} = \mu c/R$, $A = u + k(Z - v)$, and $B = k\cos\gamma + \sin\gamma$, this equation can be transformed into:

$$\sin\alpha = \frac{c}{A - Bt} \quad \text{(Eq 5.71)}$$

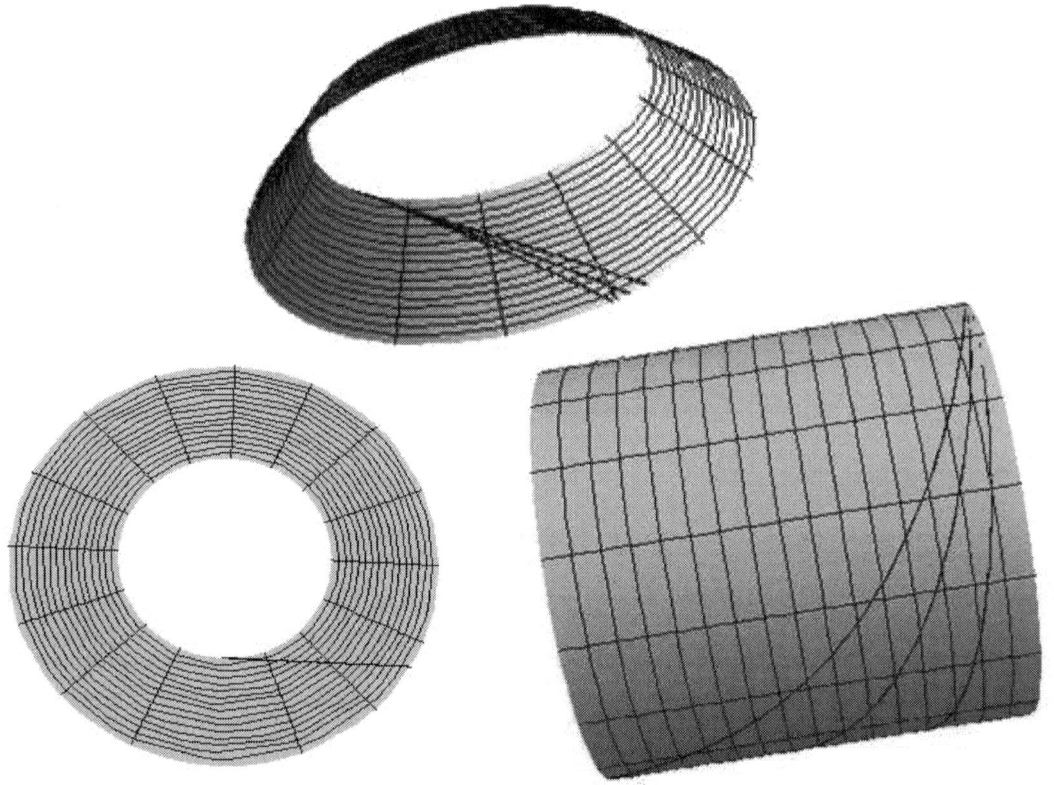

**Fig. 5.11** (Non-) geodesics on a conical shell ($\mu$ = {−0.1, 0, 0.1}), annulus ($\mu$ = 0, other values make no sense), and cylinder ($\mu$ = {0.1, 0.2, 0.3})

With Eq 5.24, the expression for $\phi'(t)$ can be derived as:

$$\phi'(t) = \frac{c}{\left[u - (\sin\gamma)t\right]\sqrt{(A - c - Bt)(A + c - Bt)}} \quad \text{(Eq 5.72)}$$

The formulation of an analytical expression for the integral of this equation is quite elaborate and will be omitted here. Nevertheless, as long as $\alpha$ does not get too close to 0 or $\pi/2$ (within the order of, say, $10^{-8}$ [rad]), the numerical procedure should not cause any problems.

## REFERENCES


5.1 G. Farin, *Curves and Surfaces for CAGD: A Practical Guide*, Morgan Kaufmann Publishers, San Francisco, CA, 2002

5.2 A. Gray, *Modern Differential Geometry of Curves and Surfaces*, CRC Press, 1993

5.3 E. Kreyszig, *Mathematical Expositions Nr. 11: Differential Geometry*, University of Toronto Press, Toronto, 1959

5.4 S. Koussios, O.K. Bergsma, and G. Mitchell, Non-Geodesic Filament Winding on Generic Shells of Revolution, *J. Mater., Part L: Des. Applic.*, Vol 219 (No. 1), 2005, p 25–35

5.5 S. Koussios and O.K. Bergsma, Friction Experiments for Filament Winding Applications, *J. Thermoplast. Compos. Mater.*, Vol 19, 2006, p 5–34

5.6 E. Kreyszig, *Advanced Engineering Mathematics*, John Wiley & Sons, Inc., New York, 1999

5.7 W.E. Boyce and R.C. DiPrima, *Elementary Differential Equations and Boundary Value Problems*, John Wiley & Sons, Inc., New York, 1986

5.8 J. van Kan, *Numerieke Wiskunde voor Technici*, Delftse Uitgevers Maatschappij, Delft, 1993 (in Dutch)

Copyright © 2011, ASM International®
All rights reserved.
www.asminternational.org

# CHAPTER 6

# Control of Filament Winding Parameters

J. Lowrie McLarty*

*If you cannot make a filament-wound product for less lifetime cost than one of metal and have it weigh less, you are doing something wrong!*

## Introduction

The hardware requirements for filament winding are described and illustrated in this text. Elementary processing equipment and advanced techniques are included. A machine with programmed motions places a resin-impregnated band of fibers onto a form or tool called the mandrel. Possible fiber band pattern orientations include helical, in-plane, basket, circumferential, or random. The mandrel or the fibers being delivered are rotated to build up the required coverage and thickness. The completed filament-wound item is cured (resin is polymerized) at room temperature, in an oven, or by other heating means, depending on the resin requirements. After cure, the wound product and mandrel are separated. An alternate part fabrication method is to use uncured filament-wound fabric as a sheet molding material. The material is removed from the mandrel, placed in a mold, and cured. Comprehensive descriptions of mandrel designs used in filament winding are presented in text and illustration. This chapter describes the equipment, defines how it is best used, and presents real-life examples.

## The Filament Winding Machine System

The basic helical filament winding machine (Fig. 6.1) consists of a group of tools, including creel, wind eye, resin impregnator, motors and controls, and heating or other curing devices (ovens, mainly) that cover a form (mandrel) with continuous resin-impregnated fibers. A creel stores continuous fibers in packages that are stationary or rotating. The creel can be stationary or alternately move from one end of the machine to the other. Dry fibers are directed through a

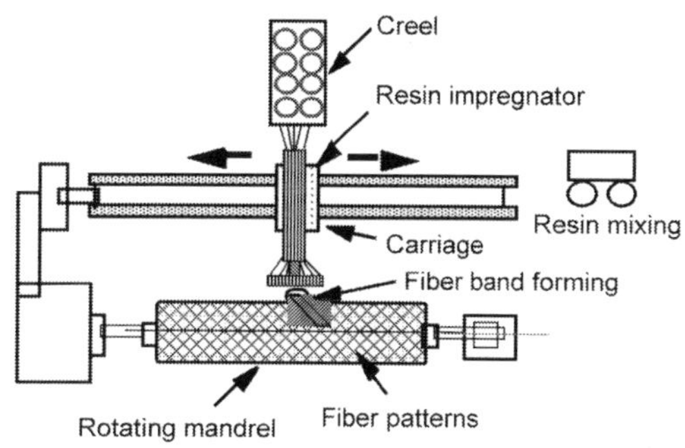


**Fig. 6.1** Basic helical horizontal filament winding machine system

*Deceased. Published with the assistance of the author's son, Douglas McLarty

resin-impregnating bath. Preimpregnated and bath-impregnated fibers are formed into a side-by-side band at the wind eye. A resin and curing agent supply with mixing device is part of the support group. The carriage containing the impregnator and band-forming device moves from one end of the machine and returns to the starting position. The carriage will place a band of resin-impregnated fibers side by side on a rotating form (called a mandrel) at angles and at progressing plus (+) and minus (–) ply patterns to cover the mandrel. The mandrel is rotated by a motor that is connected mechanically or by computer to the carriage motion. The heating device that cures the resin completes the process and includes an extraction tool that separates the cured finished component from the mandrel.

## The Basic Helical Horizontal Filament Winding Machine System

The horizontal winding machine is the backbone of the industry, and there are many different sizes and varieties of control systems, dependent on the end use (Fig. 6.2).

## Vertical Winding Machine

A vertical winding machine has an advantage for very thin-walled, small-diameter tubing, where the weight of the deposited product and the tension on the fibers could excessively deform the mandrel during winding and/or cure. However, the resin, if wet, may flow during heated cure and could end up all at one end, if proper precautions are not taken. Typical use involves prepregs rather than wet winding. The following describes one approach to fabrication via a vertical machine.

One filament winding machine made 2 to 3 in. (50 to 75 mm) diameter pipe in 20 ft (6 m) lengths at a 54.75° fiber angle (Fig. 6.3). The rotating platforms followed each other from the bottom to the top and back to the bottom. The platforms rotating in opposite directions provided the ± fiber patterns at full surface coverage. There was some migration of the resin to the lower parts of the winding. When the winding process was 5 min or less and the resin cure to B-stage was less than $\frac{1}{2}$ h, the resin movement did not allow air introduction at the top or

**Fig. 6.2** Filament winding machine. Courtesy of McLean Anderson, Inc.

increase wall thickness near the bottom that was not considered a rejection.

## In-Plane (Polar) Winder

A large version of an in-plane filament winding machine is shown in Fig. 6.4. This is the simplest design for a filament winding machine. The mandrel rotates while a fiber-delivery carriage moves around the mandrel on an angle. This motion places the fibers on a path that is called polar and is not the shortest distance from end to end on the cylindrical surface. Fibers are not liable to directly assume maximum internal pressure loads. Fibers are not in a helical path that is required for the shortest and nonslipping fiber path. For vessels up to 12 in. (305 mm) diameter, fiber slippage is usually not a problem; however, on large-diameter windings, fiber slippage can be uncontrollable. The fibers can slip free of the mandrel during cure because the resin viscosity is lower.

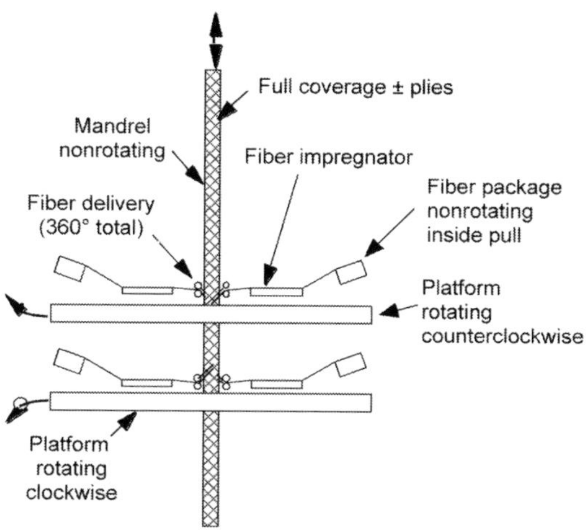


**Fig. 6.3** Vertical winding machine

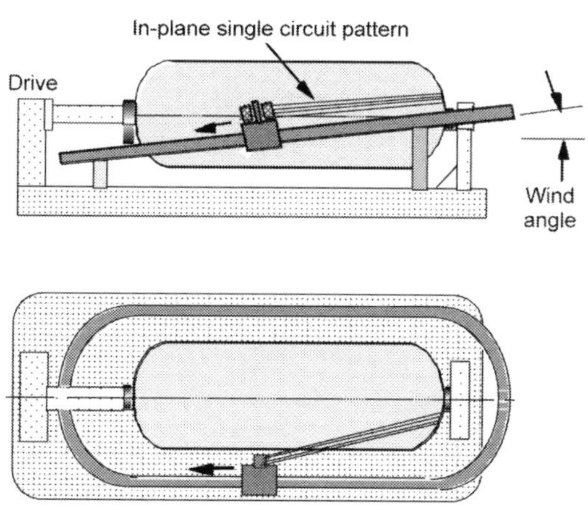


**Fig. 6.4** In-plane (polar) winder

## Creels

A creel is a device for holding the required number of roving balls or supply packages in a desired position for unwinding onto the mandrel when filament winding. There are several types: stationary and no twist, rotating, braking, and combinations thereof.

### *Stationary*

A stationary creel (Fig. 6.5) is simply a shelf from which the fiber can be pulled from the center of the bundle without tension but with twist. Including twist has consequences, as shown subsequently.

### *Twist Considerations*

The bundle of prepreg or dry fibers called tows or rovings are not twisted in their packages. It is in the fiber removal from a package

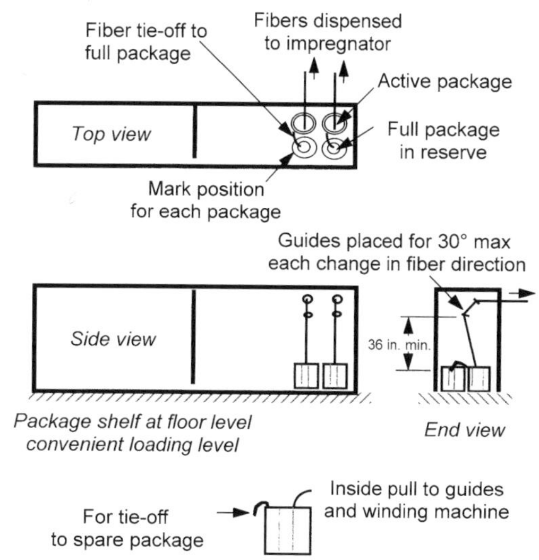


**Fig. 6.5** Stationary creel

that twisting can occur (Fig. 6.6). Fibers from inside the nonrotating package have a twist each time a diameter length of fiber is taken in a 90° direction. This creates a twist every foot or two of each fiber in the winding. Twists are also generated by fiber-delivery systems that have abrupt changes in fiber direction. The twist section of a fiber is not impregnated thoroughly in many impregnation systems. The twisted fiber does not completely flatten and can cause voids when fiber resin is at low volume fractions. Fibers do not have a twist when removed tangentially. The lowest-cost fibers in nonrotating packages are used for large filament-wound pipe and tanks, and the technology accommodates the many thousands of twists. The effect of twists is to induce fiber gap and local change in fiber shape (Fig. 6.6). Slight excess resin can fill the space beside the twist. The slight increase in height of the ply at the twist decreases the winding surface uniformity. With a twist at each diameter of the fiber package of 225-yield glass, the total number of twists in a pipe of 10 in. (255 mm) diameter and 10 ft (3 m) length with 0.030 in. (0.75 mm) wall is over 70,000. The wound surface is not as flat when fibers are twisted. The very large packages of fiber required to satisfy machine capabilities of 20,000 lb (9070 kg) per hour windings, as well as some smaller packages, can be made with a fiber twist to eliminate twist during winding.

When fibers are taken tangentially from a rotating package, there is no twist. There can be friction between the fiber being removed and the fibers remaining on the package surface. As fiber tension is increased at the creel packages, the level of fiber damage increases. Friction-causing fiber damage is reduced if the fiber being removed is pulled away from the adjacent fiber and not across its top.

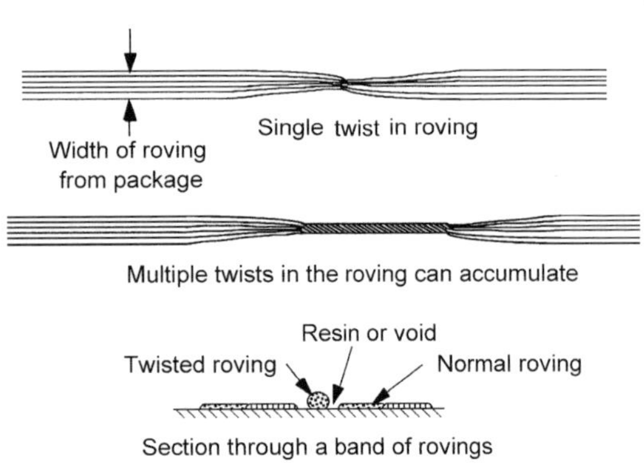


**Fig. 6.6** Twist considerations

At least one fabricator had in excess of 40 dry fiber contacts (at contact angles greater than 45°) between the creel and mandrel surface. By moving the creel closer to the winding machine, the contacts were reduced to six. As a result, there was much less fiber damage and fewer shutdowns due to broken fibers, and the winding speed was more than doubled.

### *Rotating Fiber Package Creel*

A 48-fiber package is shown in Fig. 6.7. Both carbon- and glass-fiber packages are mounted horizontally on a platform that moves with the filament winding machine carriage. Fiber tension is very low for the least fiber damage in the total fiber-delivery system. Leather straps over a steel drum can be included to create up to 2 lbf (9 N) tension for each tow or roving to eliminate fiber entanglement or catenary between creel and impregnator. For the least fiber damage, the dry fibers must remain undamaged until impregnated or beyond.

### *Rotating Creel Tensioner*

A low-cost, adjustable-force braking device for rotating fiber packages is shown in Fig. 6.8. The braking effect can be low to prevent package rotation at times of changes in demand by the machine fiber delivery at end-fiber reversal. Packages can be balanced for equal fiber tension at the band-forming device. At higher settings, the tensioner can be used to apply fiber tension for winding that has excess resin. There are a number of mechanical and electrically controlled tensioners, some with rewind capability, available from the filament winding machine vendors and independent suppliers.

**Fig. 6.7** Rotating fiber package creel

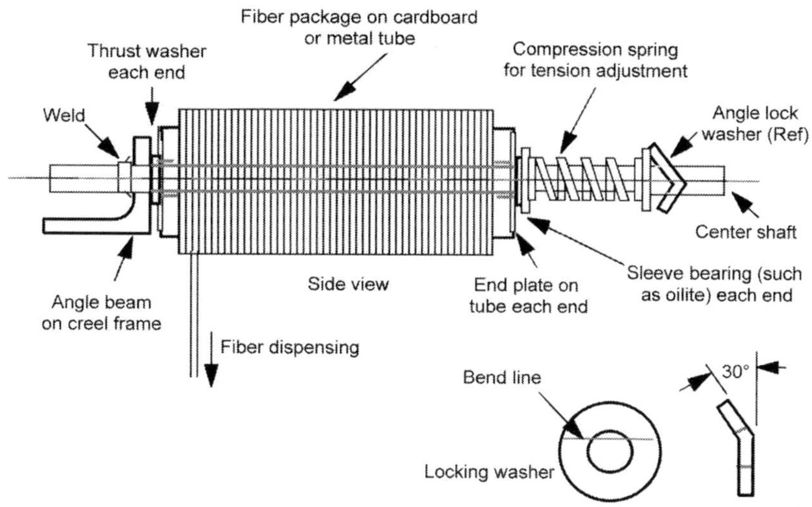


**Fig. 6.8** Rotating creel tensioner

### *Rotating Package Paths to Impregnator*

The fiber paths from creel to impregnator should have minimum guides and changes in direction (Fig. 6.9). The tension at the fiber packages is to be as low as possible. The quality of guides for the fiber path is especially important if tension is high. Fiber damage that travels through the impregnation, band forming, and delivery to the winding surface affects the filament winding performance in machine halts and fiber placement.

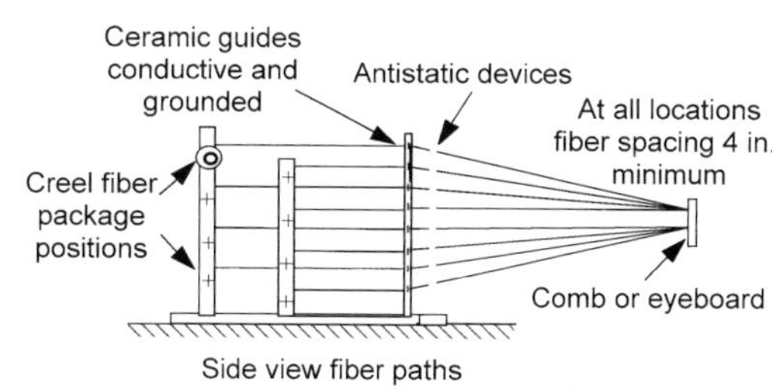


**Fig. 6.9** Rotating package paths to impregnator

### *Handling Tow for Filament Winding*

**Guides, Eyelets, and Other Orifices That Contact the Fiber.** This section addresses the delivery and handling of the tow during filament winding and was adapted from Ref 6.1 with permission. With only a small amount of attention and rudimentary engineering, carbon-fiber tow can readily be spread evenly to its maximum width or any desired intermediate width. (For a wet-wound 12k tow, nominal width is approximately 0.125 in., or 3 mm; for a 48 to 50k tow, it is 1.00 to 1.25 in., or 25 to 30 mm). Carbon fiber is the most difficult fiber to handle; thus, the good practices suggested here will be applicable to any of the other fibers when problems arise. Good practice in fiber handling will result in wider band width available, with resultant better control of layer thicknesses. Maximum integrity of fiber is achieved by having the lowest possible fiber tension until it is resin impregnated. For minimum damage, the fiber should be close to zero tension until impregnated.

The tube package must be positioned horizontally, with the tube axis perpendicular to the direction of fiber feed. Fiber should never be pulled off the end, because each revolution of feed introduces one twist. Twist is not acceptable for high-performance fibers because it decreases fiber volume and reduces some mechanical properties.

It is basically a matter of choice as to whether the fiber feed is taken from the top or bottom of the package. However, if there is zero tension or if excess slack occurs when using top-fed fiber, there is the opportunity for material to slip off the package ends and become entangled on the support shaft. Fiber feed taken from the top of the package does not pass over the adjacent fibers, reducing fiber abrasion.

**Guides, Eyelets, and Orifices That Contact the Tow.** The minimum bore diameter must not be less than 0.375 in. (9.5 mm) diameter. The entry and departure angles of the fiber through the guide should be minimized and never more than 20° from the normal (Fig. 6.10). This prevents acute fiber abrasion. The length of fiber contact within the guide should be no greater than 3⁄16 in. (4.75 mm), to limit the fiber friction.

Ceramic guides are an excellent choice, but steel may be used (as in a "pigtail," for example) only if the steel is hardened and the guide is very well polished and monitored, because grooves can be made in the steel, and the plating will deteriorate. Pigtail guides should not be of fine-wire diameter. Pigtail guides allow the winder to move the fiber from one guide to the other without cutting the fiber but should not be used with tensioned fiber. Figure 6.11 shows typical guides. Roller redirects, never guides or pigtails, should be used for prepreg.

The first guide or pick-up point should be at least 16 in. (400 mm) or three times the package diameter from the package surface (Fig. 6.12). If the guide is too close, the fiber scuffs laterally across the outer package surface, generally abrading and causing loose fiber. Additionally, the first guide should be on the horizontal centerline of the package to prevent falloff of wraps at the edges and to minimize scuffing.

Try to eliminate as many redirects (guides) as possible. The ideal fiber handling is zero touches between the creel to the wind eye, but this is generally impossible. The reason for reducing the number of redirects is that each redirect adds friction to the fiber, reducing the control that should be exercised at the creel. As guides are added, their friction is a multiplier of the previous friction. When the first guide has a fiber friction of 2 oz-f (0.5 N), the following guide has an added friction of 10 to 50%. Subsequent guides have additional friction values.

If the first guide from the package is shifted to either side of the centerline, the departure angle becomes more acute, which results in lateral slippage as it unwinds, scuffing the underlying fiber. All redirects must be generously radiused (highly preferred) or beveled.

Excessive tension should be avoided for a number of reasons. There is the possibility of increased self-abrasion over the roll surface and in guides and possible breakage of the tow. The tow may tend to embed itself in underlying windings, producing a tangle. Tension should not exceed 2.205 lb (1 kg) for 48 to 50k tows and 1 lb (454 g) for 12k tow. Tension applied to impregnated fibers may be graduated from high to low as the winding proceeds, and only

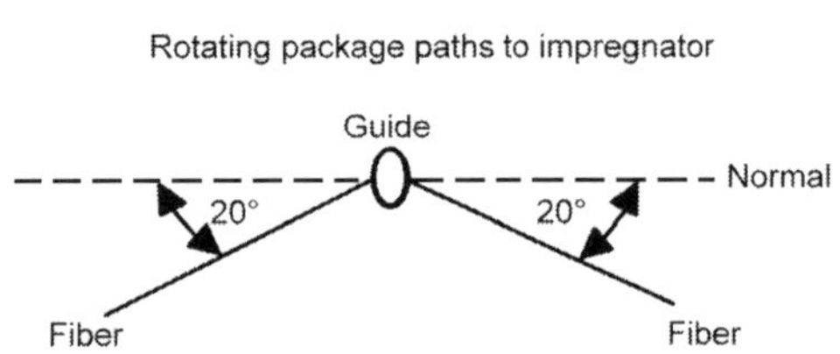


**Fig. 6.10** Fiber angles to guide

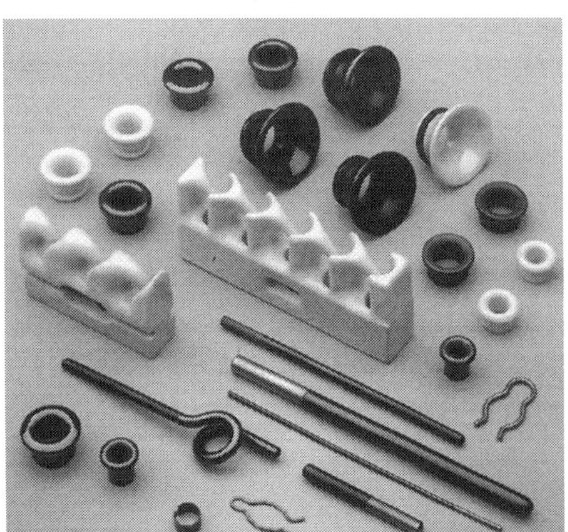

**Fig. 6.11** Typical pigtail and ceramic guides. Courtesy of Texkimp, Ltd.

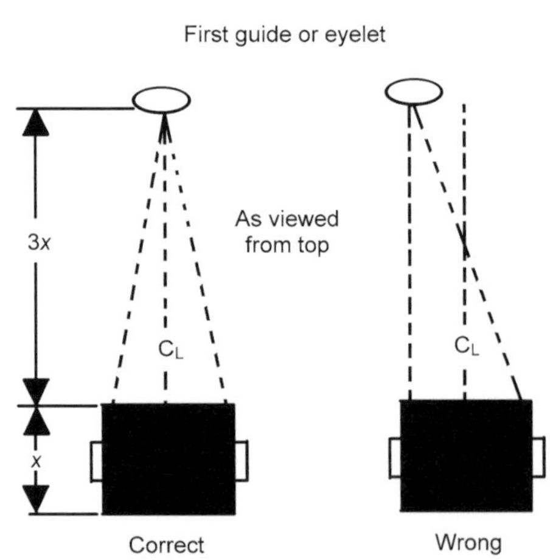


**Fig. 6.12** Guide-to-package distance

on fibers having the desired resin content. There should not be excess resin or squeegeeing necessary.

If flat sheet is used as an orifice/guide plate, choices of materials should be limited to polytetrafluoroethylene (Teflon, E.I. du Pont de Nemours and Company), ultrahigh-molecular-weight polyethylene, or nylon.

Carbon graphite is a high-modulus and therefore brittle and frangible material. There will always be some minor degree of fiber buildup at the guides and/or contact points. The essence of successful processing is what may be called "preventative housekeeping practices." In periods as short as 10 min elapsed processing time, any extraneous loose fiber buildup should be removed. If a small problem remains unattended, it will escalate very quickly into a serious problem, possibly resulting in tow breakage. Other fibers, such as glass and aramid, do not abrade or "fuzz up" as badly. Typically, loose fiber buildup problems are exacerbated with resin-wetted tow, after it exits the resin bath. Somewhat closer attention to "housekeeping" is demanded in this circumstance.

Tube-packaged fiber is provided twistfree. However, false twist can occur. Typically, if the path length from the creel to the first process point is adequate (8 to 10 ft, or 2.5 to 3 m), especially with prepreg where false twist cannot be tolerated, the false twist will fall out.

**Spreading the Fiber.** Moving the fiber over a rolling surface does not increase spreading nearly as much as passing it over static surfaces. However, in either case, the surface should be polished (5 root mean square or better—comparable to drill rod stock, which may be used).

The lower the tension level on the tow, the wider the spread width will be. Ideally, only enough tension should be placed on the tow to prevent roving overruns when winding is stopped. For wet winding, a nonreversing mechanism on the impregnation roller will prevent excessive spill problems when there are broken tows or lengthy reversals that drag the impregnated tow back over the roller.

Diameters of static bars should be no less than 0.5 in. (12 mm); 0.5 to 1.0 in. (12 to 24 mm) is a good working range. Usually, the larger the radius the fiber passes over, the greater the spread of the fiber. Fibers in a band can be spread after impregnation; concave rollers or bars can be used for narrowing the band, and convex bars or rollers for widening the band.

An "S" wrap, consisting of two static bars placed one over the other, is a very effective way to spread the fiber. The pair can be rotated as a unit to minimize or maximize the fiber contact or to provide tension in the fiber. The longer the fiber is in contact with the surface (residence time) on the static rods, the greater the spread width will be. Running speed has little or no effect on the spread width.

If multiple "S" wraps or static bars are used, path length has an effect on the spread width. A path of 2 to 4 ft (0.6 to 1.2 m) is a good working range. Beyond that, the fiber will tend to rope up or come back together (this is especially important with resin-wetted fiber). The higher the back tension on the fiber, the more critical the sweep angle becomes on unwinding. If the package is mounted with a spring, washers, and nut arrangement (for tension control) on a shaft, the package should rotate in the direction that loosens the nut (otherwise, it will tighten the nut until the package is literally seized).

## Mandrels for Filament Winding

The mandrel for filament winding can be of any material that will resist the winding tension, the heat of curing, the potential deformations caused by the weight of the product and the mandrel weight, can be made to the necessary tolerances, and has sufficient durability. They have ranged from cardboard to stainless steel. The basic requirement is that they can be removed (extracted) from the wound component unless, in some applications such as pressure vessels, the mandrel remains in the component as a liner. Some of the alternate approaches are shown subsequently.

### *Filament-Wound Mandrel*

A filament-wound vessel with fibers at ±54.75° will, under internal pressure, expand in both directions, and the wind angle will remain the same. When the angle is ±60°, the fibers move toward a ±54.75° angle. This movement decreases the diameter and increases the length of the vessel (Fig. 6.13). The filament winding mandrel uses this fact to simplify extraction of a filament-wound pipe that was wound onto it. The mandrel is fabricated by overwinding both ends of a dissolvable mandrel or a 2 to 4 lb (1 to 2 kg) compressible foam mandrel. Center shafts are used with end-pressure connections. The

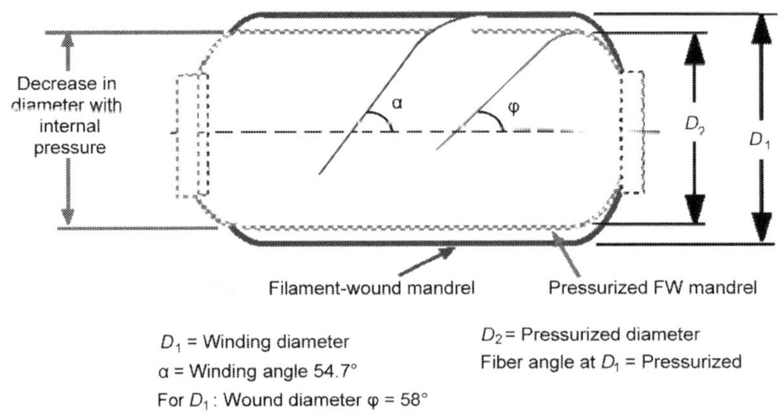


**Fig. 6.13** Filament-wound (FW) mandrel

mandrel surface can be machined by the filament winding machine using a tool postgrinder on the winder carriage. Machining does not significantly reduce the internal pressure capabilities. Plastic film can be spirally wrapped with a slight overlap as a release system. Resin for the mandrel should be cured at a temperature at least 50 °F (28 °C) greater than the cure temperature of the filament-wound product. The extraction pressure force for the mandrel is as low as 50 psi (345 kPa) and up to several hundred pounds per square inch. Trial windings and analysis can determine the most suitable mandrel wind angle. The filament winding mandrel can be repeatedly used.

## Filament Winding Process Elements

In addition to the basic winding machine and its accessories, the filament winding process involves a polymerization (curing) phase to cure (harden) the resin in the product. The heating can be on the filament winding machine using infrared radiation or in a separate hot air oven or turning rig with heat. The winding must be slowly rotated during the cure cycle until the resin does not flow. Revolving prevents the resin from flowing or introducing air at the top and a thicker wall at the bottom of the product. The cured winding is taken to an extraction station for mandrel removal.

### *Process Description*

Continuous fibers, as tows or rovings, dry or preimpregnated with resin, are pulled from fiber packages in a creel, guided through a mixed-resin impregnating system (if fibers are dry), and then formed into a flat band. A filament winding or placement machine has a fiber-band-dispensing carriage that traverses from one end of the form (mandrel) to the other (called a pass) and back again (completing a circuit). This motion alternately places a plus-angle fiber band in one direction and a minus-fiber band in the reverse direction as the form revolves. The other fiber band direction is hoops that are close to 90° to the mandrel axis. There is a motion ratio between the revolving form and the fiber bands carriage position that progresses the band of fibers until the form is covered. This is called a ± ply. This sequence of ± plies coverage is repeated as required for either thickness or number of plies. Next, the form and winding are heated or otherwise treated to cure the resin. The tooling is removed from the winding, and finishing operations, if any, are performed.

Dry fibers and wet resins in the process are in their lowest-cost form. The fibers are placed by machine, not by hand. The repeatability is great. The process can be very highly automated with excellent control. Products are always less in weight and can be less in cost than those of metal.

The filament winding and placement process is deceptively simple. Buying and operating a winding machine to make a product is not comparable to buying and operating a machine lathe to make a product. The winding machine is only one element in the chain of knowledge and equipment required to fabricate a satisfactory product. Many of the institutions making filament-wound items for testing and evaluation obtain and publish results that are not representative of properly designed, tooled, and fabricated parts. There are also industry-fabricated small and very large filament-wound structures that have needlessly failed because of resin, fiber pattern, fiber impregnation, lower-than-required resin cure temperatures, and other aspects of the process. Fabricators using incorrect resin impregnation and machines with somewhat random placement of fibers can often compete with large metal and concrete products because of the cost- and weight-savings for the customer. This allows the fabricator to waste resin and fiber to develop the thicker wall that inadequate filament winding practices require.

## Special Processes

### *Autofrettage with ±85° Fibers*

Steel or aluminum closed-end pressure vessels can be overwrapped by filament winding, cured, and then pressurized to yield the metal and place the filament winding in tension, increasing the pressure capability of pressure vessels.

The process of filament-wound autofrettage is to filament wind over a metal cylinder and then to pressurize it beyond the metal elastic limit. The metal can be aluminum, steel, or any metal that will yield. The residual stress is beneficial to the metal in an unpressurized state by placing the metal in compression. Under rated internal pressure, the metal stress is less than the stress induced by autofrettaging. The filament-wound overwrap stress is also less than the pressure developed during autofrettage. The autofrettage vessel shown in Fig. 6.14 is a steel forging with the wall thickness the same on the cylindrical section and ends. A ±85° filament-wound pattern on the cylindrical portion is sufficient for autofrettage. The thickness of metal and of filament winding is often the same. Metal cylinders with thinner ends can be filament-wound completely autofrettaged end to end. Care must be exercised to avoid cyclic fatigue in the liner and to not exceed the buckling strength in the design of the metal liner.

### *Slip-Form Continuous Winding*

Slip-form continuous winding is a filament winding procedure for producing pipe to any length. It is suitable for low-pressure applications (Fig. 6.15). The mandrel is steel tubing

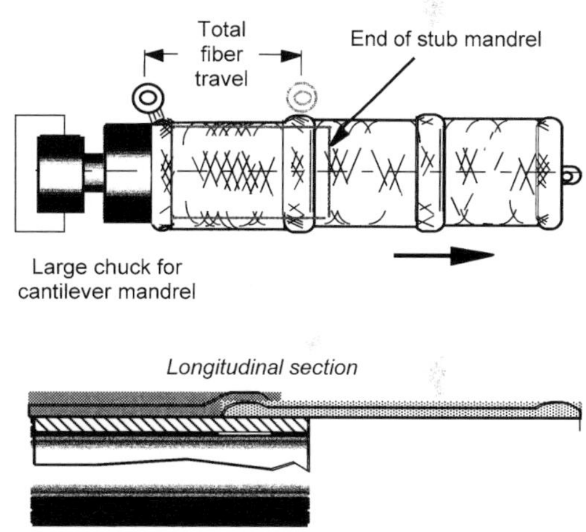


**Fig. 6.15** Slip-form continuous winding

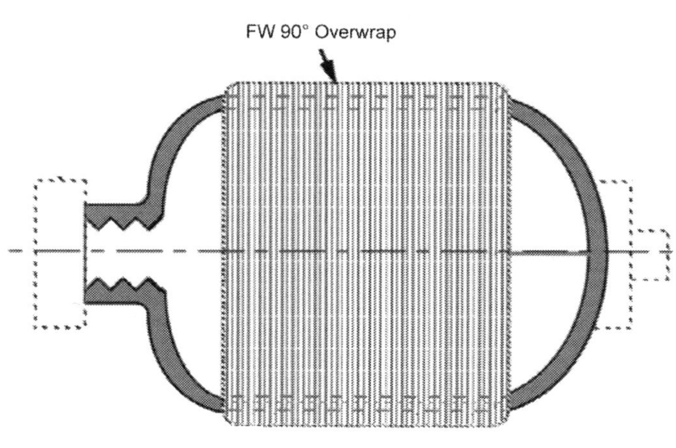


**Fig. 6.14** Autofrettage. FW, filament wound

with one end designed for cantilever mounting in the filament winding machine. The mandrel is the length of $2\frac{1}{2}$ windings. For long pipes, a wheeled support (not shown) is used under the cured portion. Extraction of the pipe is by forklift and straps around the cured end of the winding (not shown). The mandrel is covered with $\frac{1}{8}$ to $\frac{1}{4}$ in. corrugated cardboard initially and as the winding progresses. A fast curing system is used for the polyester resin and far-infrared bars for B-staging and cure. The wound part is moved when the just-completed winding is in a preliminary B-stage. The end windings are always wound into pre-B-stage resin for a primary bond. A resin-compatible adhesive is applied on the end winding for increased attachment. This technique has been used for winding large-diameter (12 ft, or 3.5 m), continuously wound silo pipe by the Drostholm fabrication process.

### *Crossing Band Thickness Reduction*

Crossing band thickness reduction is a method for reducing or maintaining the thickness of crossing fiber bands. Filament-wound isogrid lattice and basketweave structures may be required to be uniform in thickness, particularly if there is to be a continuous overwrap. This is possible when the fiber band is thick enough that fibers can be spread to reduce the thickness (Fig. 6.16) and the forming guide spacing to the forming ring is increased. The alternate is that the comb for delivery of fibers to the forming ring can be brought closer to the ring until the spread of fibers is the desired thickness. The forming ring should have an inside diameter equal to twice the new band width. This ring can also be used for forming the narrow band.

### *Filament-Wound Hole*

Holes are filament wound by passing around a pointed tool on a mandrel (Fig. 6.17). When low wind angles are combined separately with high wind angles, there is a very small area not covered by wound fibers. A doily of circumferentially wound fibers can be inserted between the wound ± plies to bridge the gap. The pointed tool is tapered and has a hole for its extraction prior to extraction of the mandrel. A washer (not shown) can be used to flatten the fiber buildup next to the pointed tool.

### *Leakage Path between Plies*

When the number of circuits per each ply is an integer, the gaps between bands of fibers present a path for leakage (Fig. 6.18). Leakage paths are most apparent with placement bands or filament-wound bands that are not allowed to have an

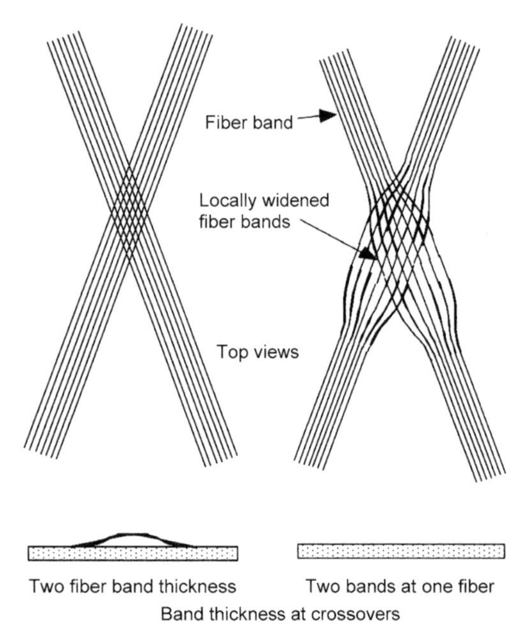


**Fig. 6.16** Crossing band thickness reduction

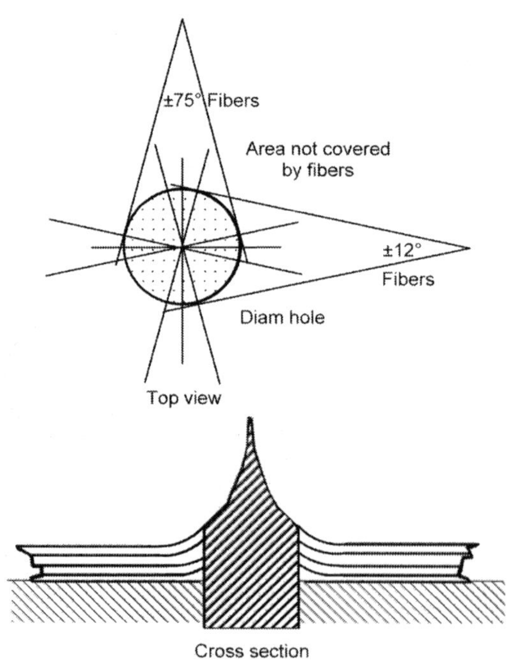


**Fig. 6.17** Filament-wound hole

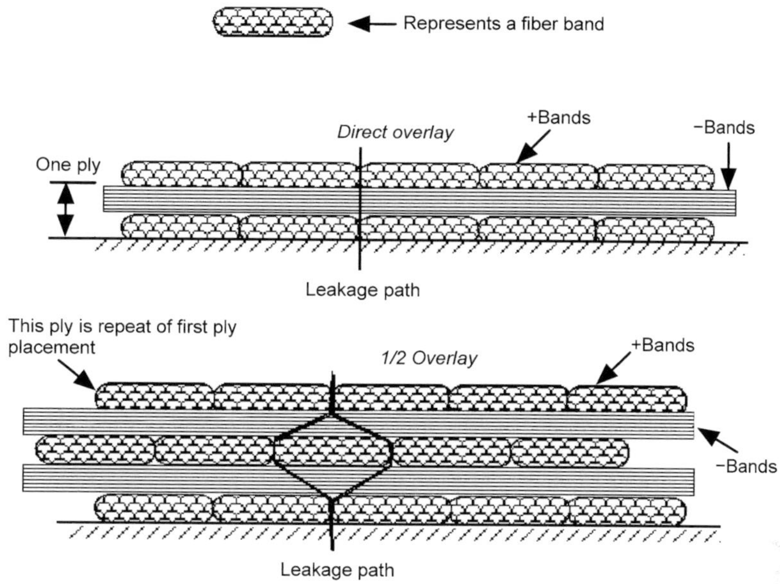


**Fig. 6.18** Leakage path between plies

overlap. When the number of circuits per two plies is an integer, the gaps exist for only one ply of bands:

Circuits per two plies = (P × 2 × Ply diameter × cos a)/Band width

### Filament Winding Progressive Cure

Curing a winding as it progresses is a means of reducing fiber tension within the ± plies that is caused by resin shrinkage. Curing filament-wound tubing increases the inside diameter and decreases the outside diameter of the uncured winding. The outer ± ply fibers are at a reduced diameter compared to when wound, and the inner plies are increased in tension. This effect is most notable and extreme for thick-wall windings. Approximately one-half the resin shrinkage occurs at B-stage. When a progressive cure schedule is used, the shrinkage effect on fiber tension/compression is reduced (Fig. 6.19). The fiber condition can also be enhanced by use of graduated fiber tension during filament winding.

### Caul Plate and Filament-Wound Surface Finish

Using a caul plate is one system for improvement of the outside finish and consolidation of a filament-wound product (Fig. 6.20). The female caul plate inner surface has the desired outer shape and surface condition for the filament-wound part. The caul plate is in thirds or half-segments of the part perimeter. An internal caul plate with silicone rubber is placed against the mandrel before initial winding. The silicone rubber in the internal caul plate provides some expansion force to move the composite component outwardly. The external caul plate provides a compression force on the completed winding during pressurization before cure.

### Clay Pipe Flange

The bell-shaped integral flange of large clay pipe was frequently damaged at the casting plant and during transport and onsite installation. Winding the composite flange on a plain clay pipe has solved that problem, with great savings in plant cost and installation (Fig. 6.21).

### Filament-Wound Boat

A car-top boat was made of completely filament-wound material, except for the balsa and brass fittings (Fig. 6.22). It is called a car top because it weighs 60 to 65 lb (27 to 29.5 kg). The mandrel was a series of boxes bolted together and removed from the overwound cockpit area, then discarded after cure. The quantity of edge-

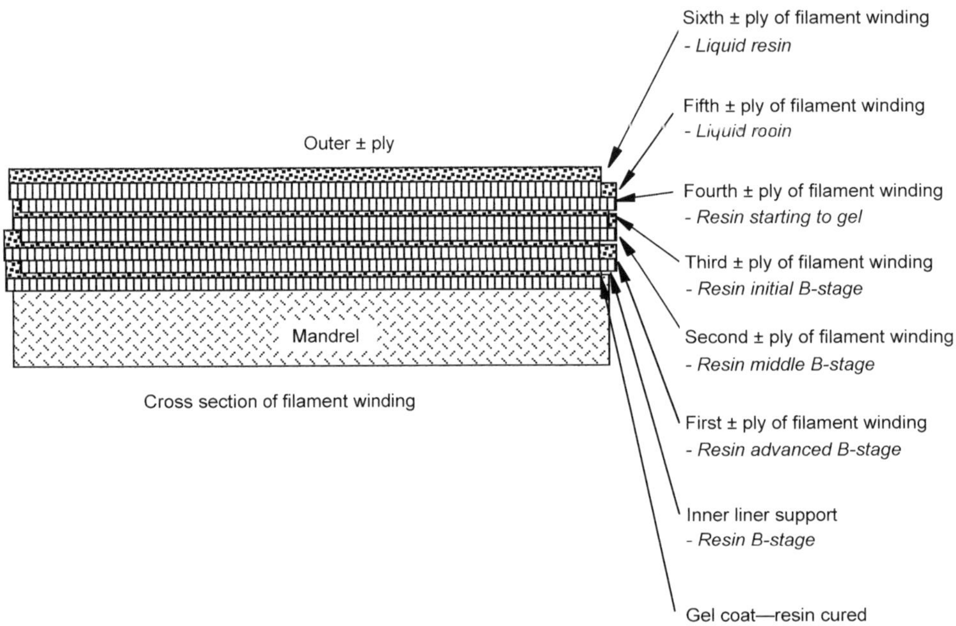


**Fig. 6.19** Filament winding progressive cure

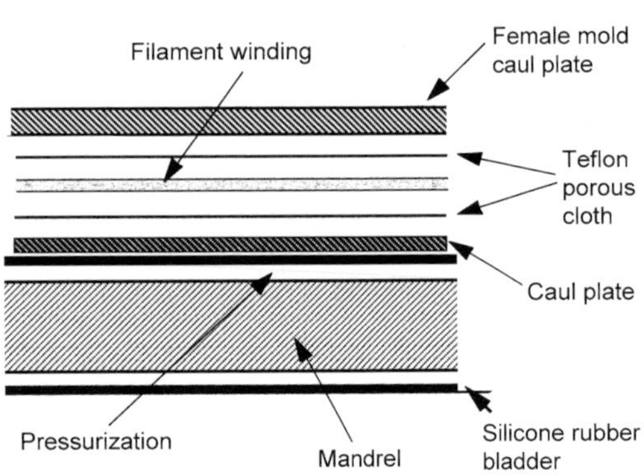


**Fig. 6.20** Caul plate and filament-wound surface finish

grained balsa within the hull allows it to float when full of water. Three seats are filament winding over balsa planks. The glass fibers and epoxy resin combination with edge balsa make a very resilient and damage-resistant vessel. Edge-grained balsa density is 7 to 9 lb/ft$^3$ (0.11 to 0.14 g/cm$^3$). Filament-wound sheet windings were used for the transom and to form the bow. The aft and forward holes for the mandrel shaft were patched with filament-wound sheet. Travel over rocks in a swift stream has left the surface without abrasion. The boat is easy to row and uses up to a 6 hp outboard.

### Filament-Wound Aircraft Fuselage

The aircraft fuselage shown in Fig. 6.23 was filament wound by Larry Ashton, presently of Rocky Mountain Composites. It has been successfully flown, and Ashton received a Lifetime Achievement Award from the American Composite Manufacturer's Association for the reduction in weight and lifetime costs for filament-wound structures.

### Filament-Wound Rib on Filament-Wound Underground Pipe

Underground large filament-wound pipe and pipe under roadways require very thick walls or the less costly reinforcing ribs at intervals along their length. The ribs are filament wound over an inverted cardboard channel wound over by a sine wave of fiber bands directly after filament winding the pipe (Fig. 6.24).

**Fig. 6.21** Clay pipe flange

**Fig. 6.22** Filament-wound boat in use

**Fig. 6.23** Filament wound composite fuselage. Courtesy of J. Winnegar, Rocky Mountain Composites, Inc

**Fig. 6.24** Filament wound rib

### *Breakaway Post*

The filament-wound component shown in Fig. 6.25 is a safety post that is as stiff as a wood post but does not shatter when bent. It can be a temporary post by fitting it in a filament-wound socket buried in concrete or earth. A locking pin at the base is optional. A urethane resin is used for resilience. The bending and failure modes can be established during production winding. Production of posts with differing characteristics is easily done with computer change-of-pattern programs. At 42 in. (1.1 m) above ground and 50 to 75 lbf (220 to 335 N) side load, the deflection with a 0.09 in. (2.3 mm) wall is 1 in. (25 mm) with 50 lbf and 0.15 in. (3.8 mm) with a wall of 0.15 in. The tube remains in one piece at failure.

### *Rapid Progressive-Cure Filament-Wound Vehicle Spring*

The spring shown in Fig. 6.26 replaces one of metal. The advantage is that the filament-wound spring is elastic and returns to shape after many load cycles. The steel spring relaxes during its life, and the spring rate declines. For the short trips of mobile homes, the steel is acceptable. The filament winding was made using a very fast twin-resin system and is therefore a progressive cure. The springs were made on a circular tool having space for four springs. Springs are not the arc of a circle. Winding was circumferential, with fibers the full width of the spring. The ring was removed from the mandrel immediately after winding was complete. The wound ring was placed in an oven for full cure, and then each spring was cut from the ring.

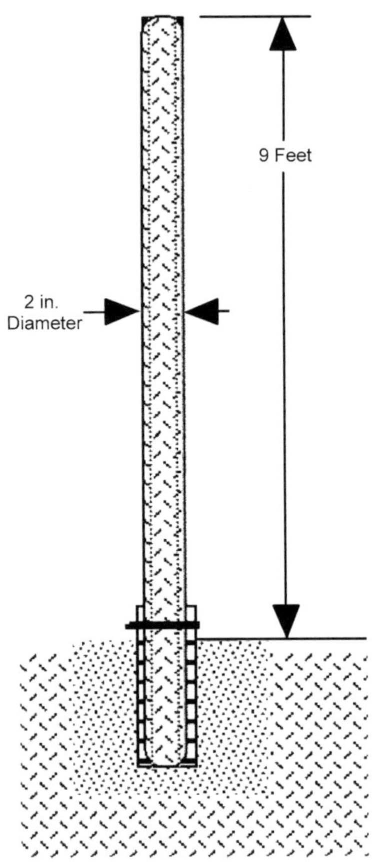


**Fig. 6.25** Breakaway post

### *Filament-Wound Disks Shock Absorber*

Filament-wound disks are a method of absorbing intermittent shocks. These shock absorbers can be used to absorb glider landing shocks and those of small road vehicles (Fig. 6.27). The glider disks are 1.25 in. (31.75 mm) diameter and 0.040 in. (1 mm) thickness. Glass fibers have been used in epoxy resin having 7 to 10% tensile elongation. Fiber volume ratio was approximately 0.65. Winding tension was 3 lbf (13 N) per 450-yield fibers. The resin attachment to the fiber and the low void percentage are important for long life. Molybdenum grease coats each side of the disk for extended life. Disks are quickly replaced. The shock-absorbing characteristics can be readily altered by the quantity of disks.

### *Filament-Wound, Small-Diameter Long Tubing*

The groove in the wood block in Fig. 6.28 has a radius equal to the finished diameter of the filament-wound tubing. The angle of the 2 to 4 lbf (9 to 18 N) tensioned, resin-impregnated fiber toward the rod tooling keeps the mandrel and the winding in the groove. The groove in the wood has a rubbed-in coating of wax. Typical tubes are 0.25 to 0.375 in. (6.4 to 9.5 mm)

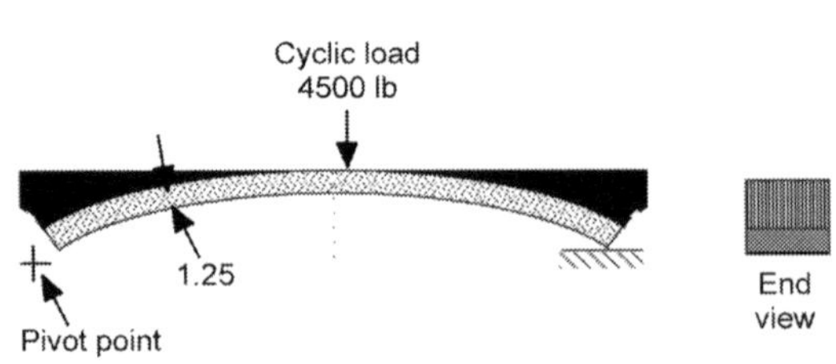


**Fig. 6.26** Rapid progressive cure filament wound vehicle spring

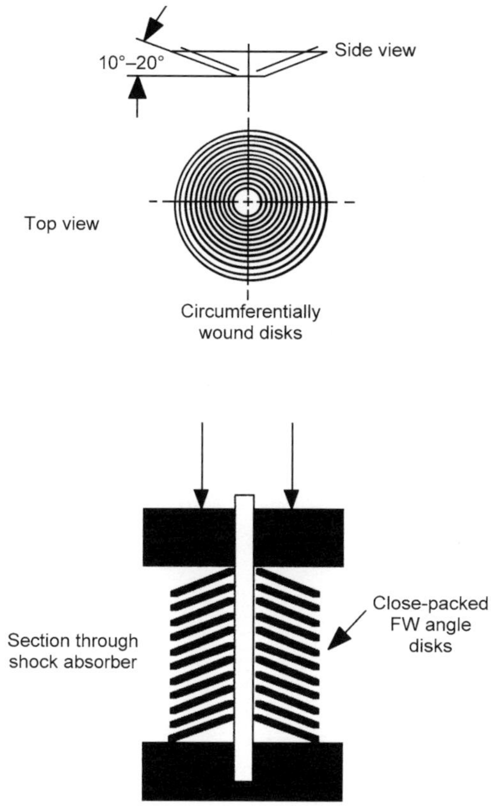


**Fig. 6.27** Filament wound disks shock absorber

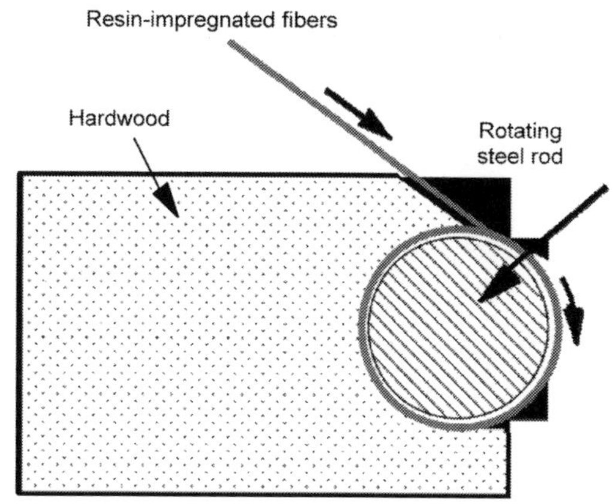


**Fig. 6.28** Small diameter long tubing

in diameter and 10 ft (3 m) in length. Curing is done on a flat screen mesh. Other wound tubes have been spirally wrapped with film in the machine.

## REFERENCE

6.1 S.T. Peters, W.D. Humphrey, and R.F. Foral, *Filament Winding; Composite Structure Fabrication*, 2nd ed., SAMPE Publishers, Covina, CA, 1998

Composite Filament Winding
S.T. Peters, editor

Copyright © 2011, ASM International®
All rights reserved.
www.asminternational.org

# CHAPTER 7

# Filament Winding Technology Learned

James Leslie, Advanced Composite Products and Technology Inc.

## Introduction

The technology of fabricating composite hardware and structures has come almost entirely from programs in which specific items were being developed or manufactured. This chapter presents specific areas of technology used in building composite parts and discusses the processes and/or articles from which the technology was derived.

## Quality Control

Many of the applications discussed involve the development or manufacture of composite hardware used in aerospace components. Fabrication for aerospace demands a high level of quality control (QC). The requirements of QC and the specific standards have paralleled the development of composite fabrication technology. Early on, the standard was Mil I 4205. This standard was improved upon and replaced by Mil Q 9858. In like manner, the currently accepted standards for both commercial and aerospace are ISO 9001:2000 and EN/JISQ/AS 9100:2004. The newest QC standard currently being used by the major aerospace corporations is Nadcap, formerly known as the National Aerospace and Defense Contractors Accreditation Program (NADCAP).

It is noteworthy that the adoption and qualification to these QC standards not only helps to assure that the parts manufactured meet drawing specifications and their intended use but also provide many additional benefits. Among these are reductions in in-house rework, reduced production costs, improved on-schedule deliveries, and improved employee morale.

## The Work Place

A very basic part of the technology of manufacturing composite hardware is the work place. No attempt is made here to describe in general the myriad of equipment, tooling, and ancillary equipment required. However, in a followup to the requirements for QC, some review of cleanliness standards is in order.

First, there are several well-defined clean-room levels that have been established to assure that parts that are manufactured will meet the requirements defined for their specific area of application. In the order of the increase in maximum number of particles per cubic meter of room air (i.e., the less clean), requirements are:

- *Class 100:* A maximum of 100 particles allowed per one cubic foot of air
- *Class 1000:* 1000 particles per cubic foot of air
- *Class 10,000:* 10,000 particles per cubic foot of air
- *GC:* Generally clean

Classes 100 and 1000 require fairly extreme procedures and equipment to attain and maintain. Class 10,000 can be attained through rigorous but not excessive work procedures and clean-room climate-control equipment. This standard meets the basic requirements for a large proportion of aerospace filament winding, bonding, and lay-up operations. Examples of Class 10,000 clean rooms for lay-up and bonding and for filament winding are shown in Fig. 7.1 and 7.2. The GC requirements can be met by simply employing good housekeeping procedures and are applicable for most commercial and much aerospace fabrication work.

## Filament-Wound Struts for the KU-3 and KU-4 Satellites

In 1988, under contract to G.E. Astro Space, graphite/epoxy struts were developed, manufactured, and flown on the KU-3 and KU-4 satellites. This work brought together and demonstrated several areas of composite fabrication technology. Figure 7.3 illustrates some of the developmental strut units produced in this program. These were the first filament-wound struts to be used on a satellite. They demonstrated that filament winding could produce tubular parts that could be qualified for space application, that these units would have higher "B-allowables," and that they could be made at a lower cost than comparable struts made by lay-up and autoclave cure methodologies. The struts were cured without external pressure, bonded into titanium end

**Fig. 7.1** Clean room

**Fig. 7.2** Clean room for winding and curing

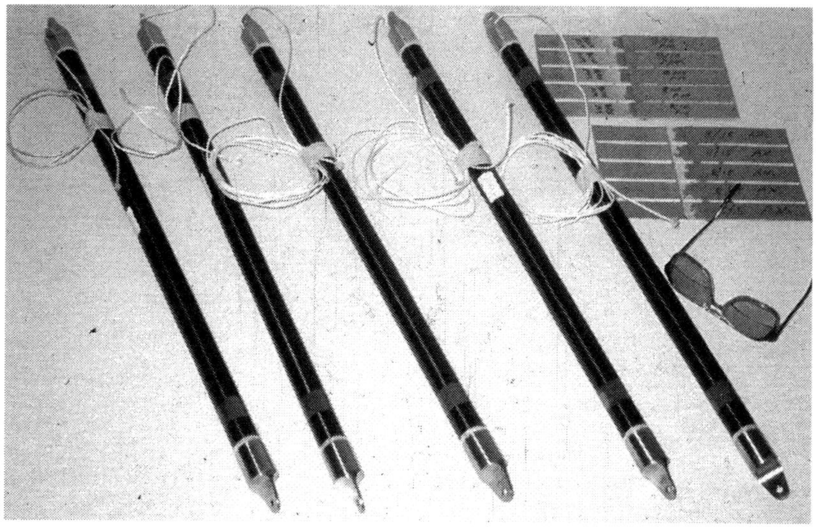

**Fig. 7.3** Filament-wound KU-3 and KU-4 satellite struts

fittings, and then successfully tested at cryogenic temperatures. The test results fully verified the design calculations: The titanium end fittings yielded prior to failure of either the graphite/epoxy tubes or the bond joints. Also as shown, and as required for most bonded aerospace structures, proof-of-bond specimens were built and tested simultaneously with each strut bonding operation.

## Off-Axial Winding of Tubular Structures

Filament winding of 0° axial plies on a tubular structure is not practical. While it is true that axial plies are necessary to achieve maximum axial properties in composite tubular structures, Fig. 7.4 demonstrates that off-axis winding can be used without serious loss of axial properties. Figure 7.4 is a typical "S-"curve, which is available in many texts on composite-oriented strength of materials. It presents the modulus of a tube as a function of the winding angle and the fiber volume. As shown at winding angles on the order of 8 to 10° off the axial, the reduction in modulus is not excessive and is often acceptable. It is noted that the off-axis winding also enhances tube stability and improves rotational properties. The same basic characteristics will be found on plots of tensile strength versus winding angles.

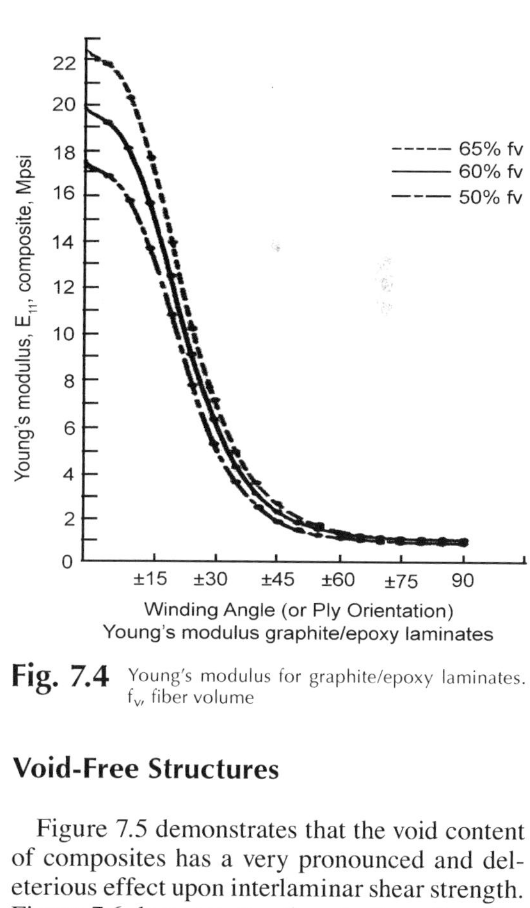


**Fig. 7.4** Young's modulus for graphite/epoxy laminates. $f_v$, fiber volume

## Void-Free Structures

Figure 7.5 demonstrates that the void content of composites has a very pronounced and deleterious effect upon interlaminar shear strength. Figure 7.6 demonstrates the winding of a void-free pressure bottle. Note that this can be a very messy procedure. First, the resin content of the structure is governed by and is very reproducible through control of resin viscosity and the tension on the fiber band. Very low, actually zero, void content is achieved by winding with excess resin. Then, the action of the fiber/resin band being wound upon the resin-rich surface forces resin up through this band, removing entrapped air. With appropriate selection of the resin, fiber tension, and ambient temperature, this technique will successfully produce void-free winding.

Figure 7.7 illustrates several different applications where this technology has been successfully used in filament-wound structures. Note that fully wound end domes, cylindrical body windings, as well as only hoop-reinforced pressure bottles are shown. The Extended Range Interceptor (ERINT) thrusters shown in Fig. 7.7 illustrate filament winding directly onto the propellant of a small rocket motor.

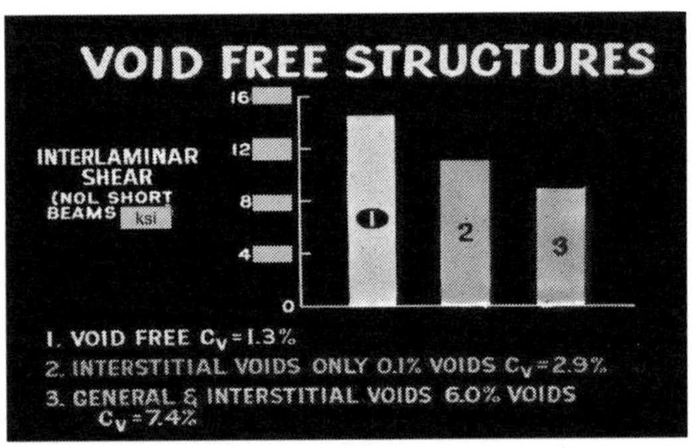


**Fig. 7.5** Void content ($C_V$) and interlaminar shear strength. 1: void free, $C_V$ = 1.3%; 2: interstitial voids only, 0.1% voids, $C_V$ = 2.9%; 3: general and interstitial voids, 6.0% voids, $C_V$ = 7.4%

## C17 Hydraulic Cylinders

Figure 7.8 shows the hydraulic cylinders as used on the C17 aircraft. This is a somewhat unique application of technology. First, the Kevlar (E.I. du Pont de Nemours and Co.) circumferential wrap on the exterior substantially reduces the weight of the cylinder. On a pressure cylinder, the radial load is approximately twice the axial load. The metal cylinder thickness is reduced to just meet the design axial load, and the overall, design-required circumferential stress is met by a Kevlar overwrap.

Second, the Kevlar is applied as a dry fiber and, after winding, is coated lightly with resin. By leaving the Kevlar dry, maximum ballistic protection is achieved. The light resin coat provides a containment sealer for the Kevlar, making the cylinder much easier to handle.

Third, the cylinders are further protected by a thin, axial graphite/epoxy lamina under the Kevlar. This laminate provides a fail-safe mechanism. Two cylinders are used in tandem. In case of ballistic damage, the graphite layer helps assure that the damaged cylinder does not interfere with the performance of the undamaged unit.

**Fig. 7.6** Winding of a void-free pressure bottle

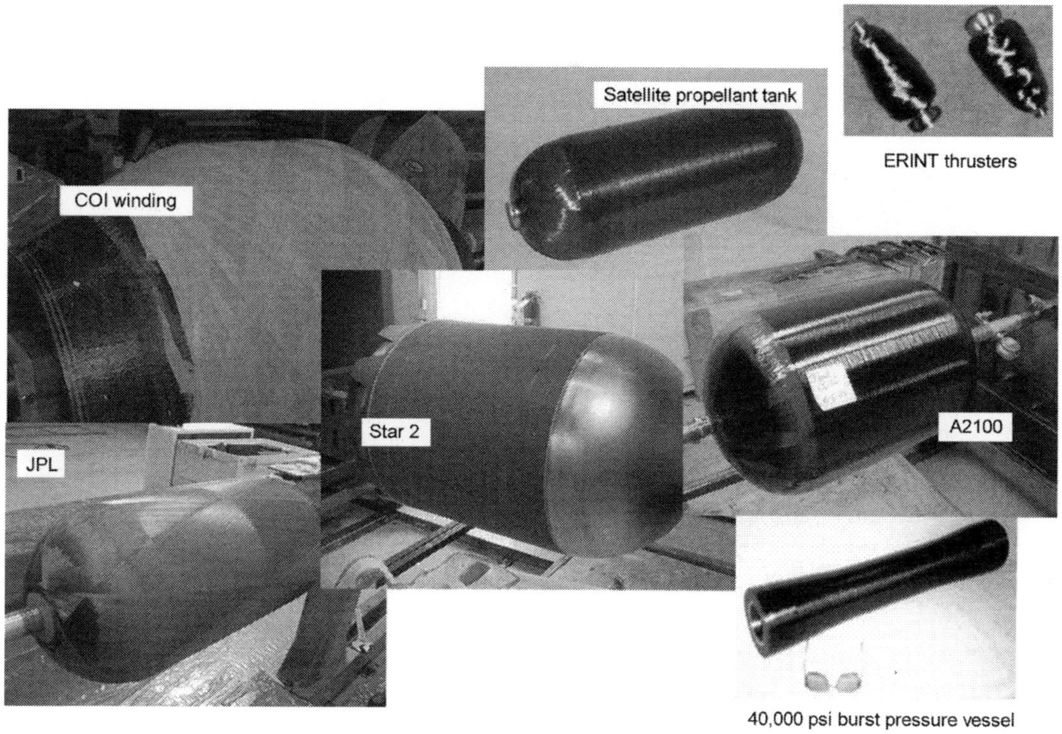


**Fig. 7.7** Pressure vessels wound with void-free techniques. ERINT, Extended Range Interceptor

**Fig. 7.8** C17 hydraulic cylinders

## Kinetic Kill Vehicles

Figure 7.9 is a picture of a full-scale demonstration model of the Space-Based Interceptor (SABIR) Flight Experiment Propulsion System. This development required two fabrication attempts. The functional requirements of the unit necessitated that the cylindrical body have close tolerance limits. Here, it was found that the thermal stresses imparted during cure at 350 °F (177 °C) resulted in unacceptable warping when the cylinder was machined. Curing at a lower temperature, to achieve an initial hardening of the epoxy, and then postcuring reduced thermal stresses to the point that the cylinder outside diameter (OD) could be machined and still remain within tolerance limits. The graphite exterior structure was formed and primarily bonded to the machined cylinder through the use of melt-out tooling.

## Composites for Cryogenic Use

A superconducting-supercollider (SSC) magnet support post, shown in Fig. 7.10, was designed and developed for the now-defunct SSC particle accelerator project. If completed, this project would have been a 54 mile (87 km) circumference cryogenic tube. The support post, as shown, would have provided insulation

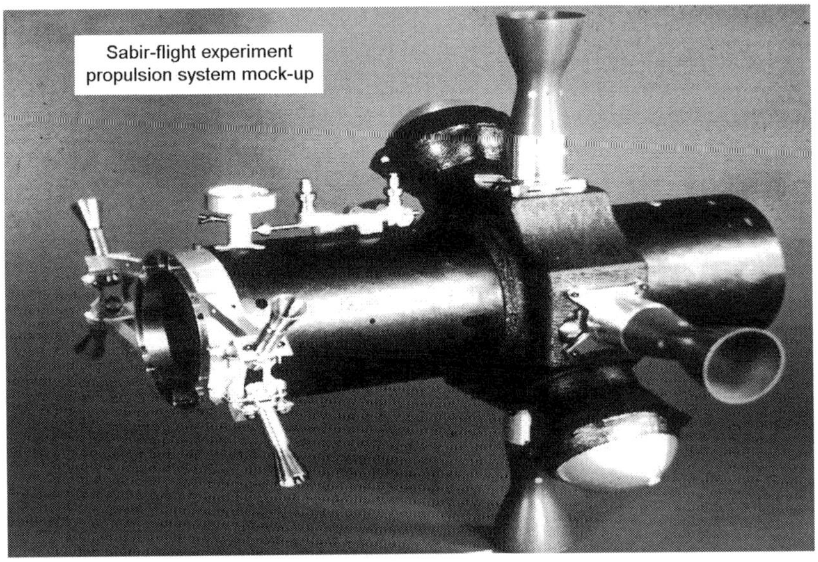


**Fig. 7.9** Full-scale demonstration model of the Space-Based Interceptor (SABIR) Flight Experiment Propulsion System

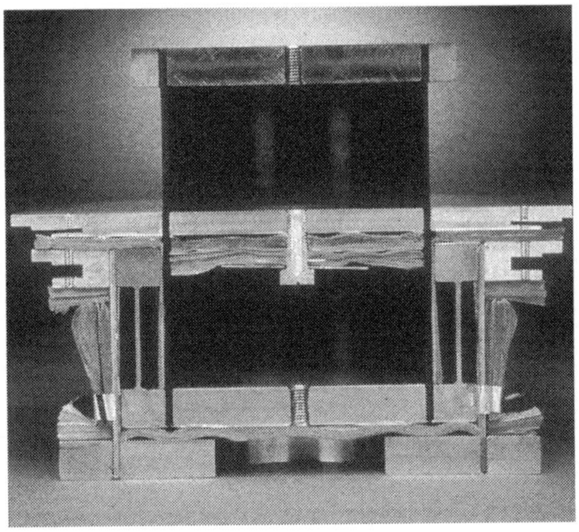

**Fig. 7.10** Superconducting-supercolliding magnet support post

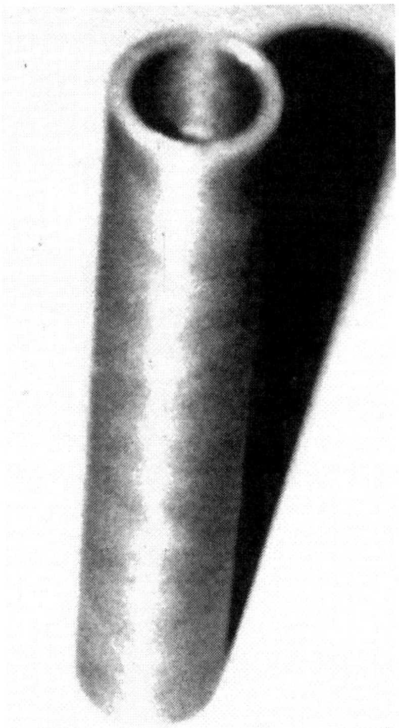

**Fig. 7.11** Quantum interference detection component

from the ambient temperature structural walls and would have been the basic structure supporting the particle beam tube. It is noteworthy that both glass/epoxy and graphite/epoxy tubes were incorporated into this design, and that the graphite tube, at the cryogenic temperatures that would be encountered, would have had a lower coefficient of heat transfer than the glass tube.

Figure 7.11 is a picture of a magnet core, superconducting quantum interference device, which is used in a quantum interference detector. This unit uses a filament-wound, ±45° copper wire core to eliminate the magnetic field that would be generated by rotation of a pure copper tube in a cryogenic atmosphere. The filament-wound copper wires are stabilized by incorporation within a filament-wound quartz glass/epoxy tube.

## Drive Shafts and Couplings

Composite torque transmission shafting is now being used in many different applications. The driving force pushing its use originates in three basic physical characteristics: vibration damping, reduced weight/lower inertia, and harmonic frequency.

As shown in Fig. 7.12, some of the current applications include commercial/industrial use in vertical pumps and cooling towers, military vehicles for both land- and water-drive systems, manned and unmanned air vehicles, wind turbines, and automobiles including racing cars.

The first successful use of a composite automotive drive shaft was by Dan Gurney's "All American Racing Team" in 1987. The shaft shown in Fig. 7.13 has over 10,000 racing miles and was used by the Gurney team to win the GTO championship in 1987. In 1988, the Chevy Beretta used a graphite drive shaft in winning the GTU title.

In both cases, the reason for using graphite drive shafts was a significant reduction in vibration. Gurney's team showed that, after installation of the graphite shaft, they could increase their "allowable red line" from 7000 to 8000 rpm, thus providing a full 12% higher speed in every gear.

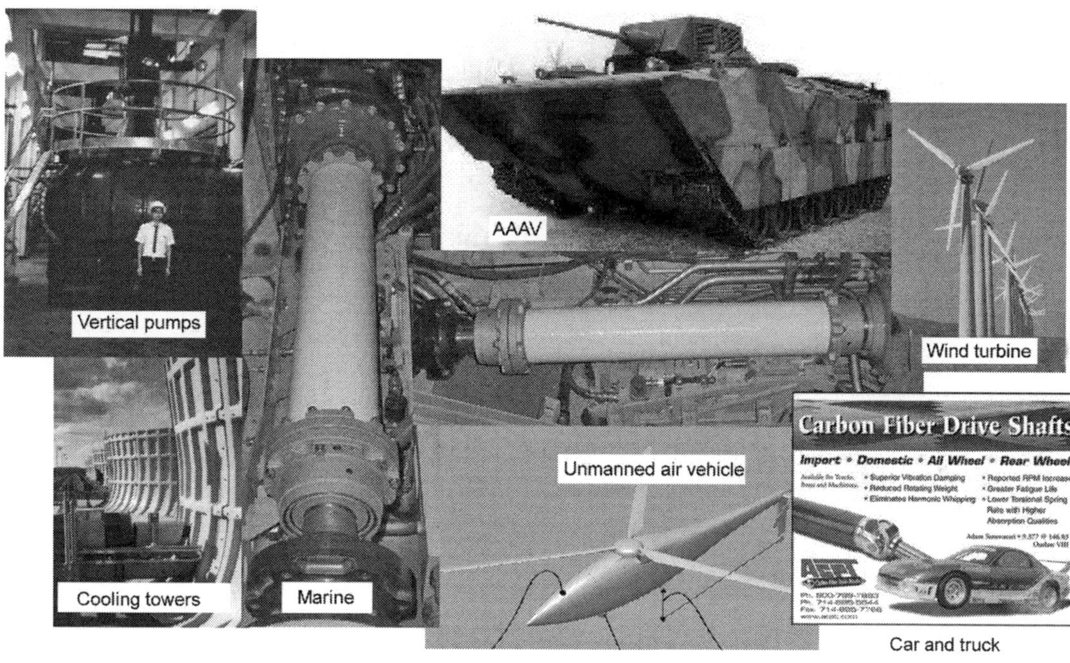


**Fig. 7.12** Drive shafts and couplings. AAAV, advanced amphibious assault vehicle

> **GRAND TOURING SPORTS CARS**
>
> The first Grand Touring (GT) classes of sports car racing were introduced in 1971 and featured six races. The GT classes are similar to the European classes of Group 2 and Group 4 racing. The U.S. classes were:
>
> - *GTO cars:* Grand Touring-type cars with an engine of 2.5 L displacement or more. This category was dominated by Corvettes, then by Shelby Mustangs, and then the various factory teams consisting of Mercury Cougars, Mazda RX-7s, Toyota Celicas, and finally the Nissan 300ZX.
> - *GTU cars:* Grand Touring-type cars with engines of 2.5 L displacement or less. This series was dominated by the Porsche 914-6, then the Datsun 280ZX, and finally by the SA22 Mazda RX-7s (1978–1985) through the end of the 1980s.
>
> For more information about GT races see Ref 7.2.

Fig. 7.13 Dan Gurney's drive shaft in 1987

Fig. 7.14 An 80,000 lb waste hauler. The challenge is to develop a composite drive shaft.

As shown in Fig. 7.14 to 7.16, an on-the-job demonstration of graphite drive shafts has proven that they can be used in truck applications. To date, however, although tested and approved by several major truck-building corporations, none are being used by the trucking industry. It is believed that the vibration damping could result in a significant reduction in downtime, but the industry has set a dollar per pound of weight-saving limit, and the composite shafts do not meet this limit at this time.

Figure 7.17 shows a shipment of developmental drive shafts for the U.S. Expeditionary Fighting Vehicle. This vehicle has both land- and water-drive systems, and graphite shafts will be used for both.

Fig. 7.15 Existing two-piece steel shaft

Fig. 7.16 Single-piece composite solution for Mack truck

Representative composite rollers and idlers are shown in Fig. 7.18. All are used for many different applications. A wide range of capabilities is provided by composite rollers and idlers in the converting industry, where the stiffness, light weight/low inertia, and overall versatility greatly assist in allowing large, lightweight, and fragile (such as toilet paper) sheet goods to be converted (cut) into commercially useful widths.

Fig. 7.18 Representative composite rollers and idlers. Surfaces can be customized with metallic, carbide, rubber, or conductive coatings or modified to have tailored finishes such as rough/gritty, smooth, or grooved. Rollers can be designed for properties such as corrosion resistance, specified thermal expansion coefficient, and natural damping properties.

Fig. 7.17 Expeditionary Fighting Vehicle propulsion couplings

Typically, the very large rolls of these films (as they are manufactured) are placed in a slitting line that is run continuously at speeds up to 3000 ft/min (914 m/min). The graphite rollers and idlers greatly reduce the operating cost and increase the effectiveness and reliability of these converting or slitting lines.

## Composite Drill Pipe

In 1998, the United States Department of Energy awarded Advanced Composite Products and Technology Inc. contract DE-FC26-99FT40262 for "Development and Manufacture of Cost-Effective Composite Drill Pipe" (CDP). As a result of this contract, full-size 6 in. (15 cm) OD, 30 ft (9 m) long pipe sections have been designed, developed, and statically tested. These units meet the mechanical specifications defined through discussions with representatives of the drilling industry (Ref 7.1).

During the course of this program, a smaller, flexible pipe was designed, developed, tested, and is being used to revitalize oil and gas wells previously thought to be "watered out," that is, no longer economically productive. Tables 7.1 and 7.2 present the physical/mechanical properties of these CDPs. As shown, two designs of the smaller-diameter pipe are available. Short radius (SR-CDP) is for use in drilling wells through a 60 ft (18 m) radius, and the ultrashort radius (USR-CDP) is for drilling through a 30 ft (9 m) radius (Ref 7.3).

Figure 7.19 is a picture of two sections of the larger pipe. The major technology development in this effort was the design of the metal-to-composite end fitting. As shown in Table 7.1 the large pipe is capable of 450 ksi (3100 MPa), and Table 7.2 shows a 75 and 65 ksi (515 and 450 MPa) capability, respectively, for the smaller pipe. Figure 7.20 is a picture of a 30 ft (18 m) section of the small pipe and illustrates its comparative light weight. Table 7.3 compares the weights of the composite drill pipes to corresponding metal pipe.

Figure 7.21 illustrates the flexibility of the smaller pipe, while Fig. 7.22 to 7.24 show a shipment of SR-CDP and its assembly into a drilling platform. Figure 7.25 demonstrates the use of the SR-CDP and USR-CDP in drilling horizontally into an oil- and/or gas-bearing strata. By going back down into the old well bore and then turning sharply into the petroleum-bearing layer, a much greater area is opened to the drilling operation.

Figures 7.24 and 7.26 show that two of the previously believed weaknesses of composites have been overcome. First, it was believed that CDP would be too fragile to be handled in an oil field. As illustrated, special precautions or handling fixtures are not needed. Second, it was assumed

**Table 7.1 Extended-reach/deep water product data sheet**

| Mechanical specification | | |
|---|---|---|
| Bending stiffness | EI | $180 \times 10^6$ lb-in.$^2$ |
| Torsional stiffness | GJ | $115 \times 10^6$ lb-in.$^2$ |
| Axial stiffness | EA | $33.4 \times 10^6$ lb |
| Rated tension load | P | 450,000 lb |
| Rated torsion load | T | 25,000 ft-lb |
| Rated compression load | $P_c$ | 250,000 lb |
| Rated internal pressure | $P_i$ | 9,000 psi |
| Max. service temperature | °F (°C) | 350 °F (175 °C) |
| **Design Specifications** | | |
| Tube inside diameter | ID | 5 in. |
| Tube outside diameter | OD | 6 in. |
| Length (pin-to-box) | ft | 30 |
| Centralizers | Optional | ... |
| Weight (30 ft sections) | lb | 375 |
| **Connection specifications** | | |
| Pin/box diameter | OD | 7 in. |
| Bore | ID | 4½ in. |
| Thread | IF | NC56 |

Source: Ref 2

**Table 7.2 Specifications for short-radius composite drill pipe (SR-CDP) and ultrashort-radius composite drill pipe (USR-CDP)**

| Characteristic | ACPT SR-CDP | ACPT USR-CDP |
|---|---|---|
| Bending modulus, msi ($10^6$ psi) | 4.9 | 2.4 |
| Shear modulus, msi ($10^6$ psi) | 3.1 | 2.5 |
| Tensile ultimate, lb | 75,000 | 65,000 |
| Tensile operating, lb | 25,000 | 25,000 |
| Compression ultimate, lb | 100,000 | 65,000 |
| Compression operating, lb | 50,000 | 25,000 |
| Torque ultimate, lb-ft | 6,000 | 6,000 |
| Torque operating, lb-ft | 2,000 | 3,000 |
| Internal pressure ultimate, psi | 2,000 | 3,000 |
| Internal pressure operating, psi | 1,000 | 1,500 |
| Collapse ultimate, psi | 2,000 | 3,000 |
| Collapse operating, psi | 1,000 | 1,000 |
| Temperature, °F (°C) | 325 (163) | 325 (163) |

Source: Ref 2

**Fig. 7.19** Composite drill pipe, 6 in. Source: Ref 7.3

**Fig. 7.20** Short-radius composite drill pipe weighing 92 lb (42 kg). The same part made of steel weighed 157 lb (71 kg). Source: Ref 7.1

that composites could not withstand the hostile environment encountered "down-well." After 160,000 cycles in a drilling operation, virtually no wear or abrasion was present in the drill pipe sections shown.

## REFERENCES

7.1 J.C. Leslie, J.R. Jean, H. Neubert, L. Truong, and J.T. Heard, "Development and Manufacture of Cost Effective Drill Pipe," 2002 Annual Technical Progress

**Table 7.3 Drill pipe weight comparisons**

| Drill pipe with 7 in. OD joints | Steel | Composite |
|---|---|---|
| Tube ID | $5\frac{5}{8}$ in. | 5.0 in. |
| Tube OD | 6.0 in. | 6.0 in. |
| Weight of a 30 ft section | 826 lb | 358 lb (3.9 lb Gr./ft) |
| **Drill pipe with $3\frac{3}{8}$ in. OD joints** | | |
| Tube ID | 2.0 in. | $1\frac{5}{8}$ in. |
| Tube OD | $3\frac{3}{8}$ in. | 2.56 in. |
| Weight of a 30 ft section | 157 lb | 76.4 lb (SR-CDP)<br>100 lb (USR-CDP) |

OD, outside diameter; ID, inside diameter; SR-CDP, short-radius composite drill pipe; USR-CDP, ultrashort-radius composite drill pipe

**Fig. 7.21** Short-radius composite drill pipe. Source: Ref 7.1

**Fig. 7.22** Short-radius composite drill pipe

Report, DOE Cooperative Agreement DE-FC26-99FT40262, Sept 20, 2002

7.2 International Motor Sports Association at www.imsaracing.net.

7.3 J.C. Leslie, J.C. Leslie, Jr., J.T. Heard, and L. Truong, Advances in Composite Drilling Components Lead to Evaluation for Critical E&P Applications, *Drill. Contract.*, July/Aug 2007, p 78–82

**Fig. 7.23** Composite drill pipe being added to a drill string

**Fig. 7.24** Rambler Energy Services adding a joint of short-radius composite drill pipe to the drill string on French #4. Source: Ref 7.3

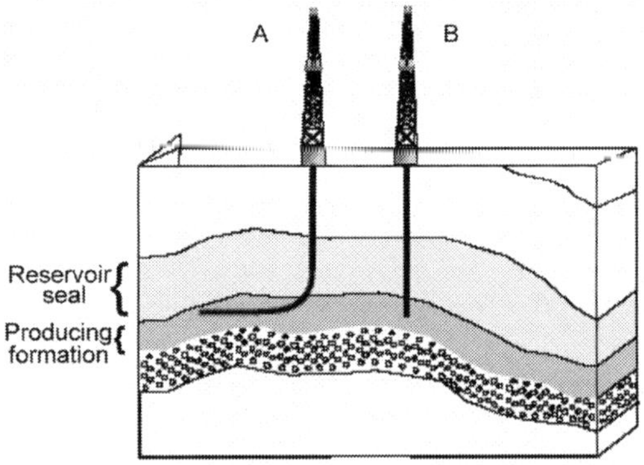


**Fig. 7.25** Use of short- and ultrashort-radius composite drill pipes in drilling horizontally into an oil- and/or gas-bearing strata. A, horizontal well; B, vertical well. Source: Energy Information Administration, Office of Oil and Gas

**Fig. 7.26** The short- and ultrashort-radius composite drill pipes exhibit little to no signs of wear after 160,000 cycles.

Composite Filament Winding
S.T. Peters, editor

Copyright © 2011, ASM International®
All rights reserved.
www.asminternational.org

# CHAPTER 8

# Static Mechanical Tests for Filament-Wound Composites

S.T. Peters, Process Research, Consultants

## Introduction

This is the sixth effort (Ref 8.1–8.5) to chronicle the extent of test techniques for filament-wound laminated composites and the materials and processes that support fabrication. This chapter follows the general approach introduced by Foral in Ref 8.2. Many of the official procedures, since the last article, have evolved, been withdrawn, or cancelled. Some of the text and figures have appeared earlier, in Ref 8.3, and are reproduced here with permission. Several fibers and resins are mentioned in this chapter and others. See the terms *fiber* and *resin* in the Glossary of this book for explanations.

The objective of mechanical testing of an engineered material is to provide data necessary for analysis, design, and fabrication of structural components using the material. The testing of filament-wound composite materials offers unique challenges because of the special characteristics of composites. Factors not considered important in metals testing are very important in testing composites. For example, composites are anisotropic, with properties that depend on the direction in which they are measured. Strain and load rate must be carefully monitored, specimen conditioning (drying, storage, etc.) can significantly affect results, and even humidity conditions at the time of specimen fabrication can appreciably affect the material. Fiber content and void content, which can also vary with manufacturing conditions, have important effects on material properties. All personnel involved in the generation and use of test results, from fabrication of the material to final data interpretation, must be familiar with these factors and their influence on test results. References 8.6 and 8.7 provide reference monographs on composites testing, and there is a wealth of test methods to evaluate composite properties using flat panels.

Filament winding is not a manufacturing method that can easily produce flat components or test specimens. There is usually a need to provide some test data to designers in addition to the modulii and ultimates derived from micromechanics data. Designers have been consistently more confident with data derived from test specimens that parallel the configurations of the laminate layup. These are the test specimens that support confidence in the filament winding manufacturing process. Usually, it is not possible to provide A- or B-basis allowables so that the filament-wound structure can be confidently designed. There are data for aircraft designers, B-basis allowables, that are provided for unidirectional or woven prepreg laminates which are hand or automatically laid up and cured and compacted in an autoclave, but these data are not generally available for the filament-wound component, even in MIL-HDBK-17C (Ref 8.8) or from vendors and Advanced General Aviation Transport Experiments (e.g., Ref 8.9). The problem is resolved, to some extent, with the selection of easy-to-use test techniques for subscale testing.

The purpose of this chapter is to describe suitable mechanical test techniques for characterizing laminated composite materials. The approach is to provide recommended techniques, based on consensus opinions of fabricators and users of filament-wound composites, and to provide a nonexhaustive survey of available techniques that have been used successfully in the field.

Some of the procedures described are standard techniques; others have potential to become standard techniques. As with any testing methodologies, but especially because of the complexities involved, the need for standardization of composites testing methods is very important. Current leaders in the standardization effort are:

- ASTM International, especially through its Subcommittee D-30 on high-modulus fibers and their composites and its Subcommittee F17.40 on plastic piping systems (Ref 8.10)
- MIL-HDBK-17 (*Military Handbook-17*), C Revision, Technomic Publishing and Materials Sciences Corporation (Ref 8.8)
- International Standards Organization (ISO), Committee TC 58 on gas cylinders and Committee TC 138 on plastic pipe, fittings, and valves for the transport of fluids (Ref 8.11)

In addition, numerous military specifications, as well as special committees of several technical societies, including the American Society of Mechanical Engineers, the Society of Engineering Mechanics, and others, are instrumental in this standardization effort.

The value of standard test techniques is that the standard represents a consensus approach, which, if followed, provides comparable, consistent, and useful results. In addition, the user can refer to the standard and know exactly the procedures that were followed in generating the data. The most accessible standards are those published by ASTM International (Ref 8.9); these are available in many libraries. The index of this multivolume set provides an extensive subject listing and an alphanumeric listing of all the standards. This index and the accompanying standards are a most important information resource for individuals setting up a composites testing program. Unfortunately, standard tests are not available for, or are not pertinent to, the needs of many filament winding applications. When no standard exists, special test techniques must be developed or the existing standards modified. In this case, the user of the data must maintain very good communication with those who generate the data. This includes the individuals fabricating the material, machining and conditioning the specimens, as well as those running the test and interpreting the data. Everyone must be aware of the interactions of all these functions on the validity of the information generated.

Data for use in design require statistical significance with a known confidence level. MIL-HDBK-17C provides a guide concerning the number and type of tests sufficient to establish statistically based material properties. MIL-HDBK-17C contains somewhat enhanced statistical treatment approaches. The three classes of allowables pertinent to current usage of filament-wound composites are:

- *A-basis allowable:* The value above which 99% of the population of values is expected to fall, with a confidence of 95%
- *B-basis allowable:* The value above which 90% of the population of values is expected to fall, with a confidence of 95%
- *S-basis allowable:* The value that is usually the specified minimum value of the appropriate government specification

For most flight-worthy composites, material properties are usually required to be either A- or B-basis allowables.

## Constituent Properties

Table 8.1 provides a listing of some ASTM International standards to determine mechanical properties of composite constituents. MIL-HDBK-17C also provides a comprehensive list of tests to evaluate reinforcement fibers, resin material, and prepregs. Aside from internal research and development (IR&D) activities, most testing of fiber and matrix materials by the user is performed as a quality-assurance procedure. For example, strand tensile strength tests, periodically performed on incoming roving, provide a record of roving quality that may be important if anomalies develop in subsequent processing or product performance. Reference 8.8 also provides a comprehensive list of tests to evaluate reinforcement fibers, resin material, and prepregs.

## Fiber Properties

ASTM D 4018 provides a method for determining strength and stiffness of the impregnated carbon-fiber tows. This information is

## Table 8.1 ASTM International standards for filament-wound composite testing

| Standard number | Standard title |
|---|---|
| **Some ASTM support specifications** | |
| ASTM C 613/C 613M-97(2008) | Standard Test Method for Constituent Content of Composite Prepreg by Soxhlet Extraction |
| ASTM D 618-08 | Standard Practice for Conditioning Plastics for Testing |
| ASTM D 792-08 | Standard Test Methods for Density and Specific Gravity (Relative Density) of Plastics by Displacement |
| ASTM D 883-08 | Standard Terminology Relating to Plastics |
| ASTM D 1505-03 | Standard Test Method for Density of Plastics by the Density-Gradient Technique |
| ASTM D 2291-09 | Standard Practice for Fabrication of Ring Test Specimens for Glass-Resin Composites |
| ASTM D 2343-09 | Standard Test Method for Tensile Properties of Glass-Fiber Strands, Yarns, and Rovings Used in Reinforced Plastics |
| ASTM D 2584-08 | Standard Test Method for Ignition Loss of Cured Reinforced Resins |
| ASTM D 2734-09 | Standard Test Methods for Void Content of Reinforced Plastics |
| ASTM D 3171-09 | Standard Test Methods for Constituent Content of Composite Materials |
| ASTM D 3529/D 3529M-97(2008) | Standard Test Method for Matrix Solids Content and Matrix Content of Composite Prepreg |
| ASTM D 3530/D 3530M-97(2008) | Standard Test Method for Volatiles Content of Composite Material Prepreg |
| ASTM D 3531-99(2009) | Standard Test Method for Resin Flow of Carbon Fiber-Epoxy Prepreg |
| ASTM D 4018-99(2008) | Standard Test Methods for Properties of Continuous-Filament Carbon and Graphite-Fiber Tows |
| ASTM D 5687/D 5687M-95(2007) | Standard Guide for Preparation of Flat Composite Panels with Processing Guidelines for Specimen Preparation |
| ASTM D 5947-06 | Standard Test Methods for Physical Dimensions of Solid Plastics Specimens |
| **Some ASTM standards for composite tensile properties** | |
| ASTM D 638-08 | Standard Test Method for Tensile Properties of Plastics |
| ASTM D 1599-99(2005) | Standard Test Method for Resistance to Short-Time Hydraulic Pressure of Plastic Pipe, Tubing, and Fittings |
| ASTM D 2105-01(2007)e1 | Standard Test Method for Longitudinal Tensile Properties of "Fiberglass" (Glass-Fiber-Reinforced Thermosetting-Resin) Pipe and Tube |
| ASTM D 2290-08 | Standard Test Method for Apparent Hoop Tensile Strength of Plastic or Reinforced Plastic Pipe by Split-Disk Method |
| ASTM D 3039/D 3039M-08 | Standard Test Method for Tensile Properties of Polymer-Matrix Composite Materials |
| ASTM D 5083-08 | Standard Test Method for Tensile Properties of Reinforced Thermosetting Plastics Using Straight-Sided Specimens |
| ASTM D 5450/D 5450M-93(2006) | Standard Test Method for Transverse Tensile Properties of Hoop-Wound Polymer-Matrix Composite Cylinders |
| ASTM D 7291/D 7291M-07 | Standard Test Method for Through-Thickness "Flatwise" Tensile Strength and Elastic Modulus of a Fiber-Reinforced Polymer-Matrix Composite Material |
| **Some ASTM standards for composite compressive properties** | |
| ASTM D 695-08 | Standard Test Method for Compressive Properties of Rigid Plastics |
| ASTM D 3410/D 3410M-03(2008) | Standard Test Method for Compressive Properties of Polymer-Matrix Composite Materials with Unsupported Gage Section by Shear Loading |
| ASTM D 5449/D 5449M-93(2006) | Standard Test Method for Transverse Compressive Properties of Hoop-Wound Polymer-Matrix Composite Cylinders |
| ASTM D 6415/D 6415M-06ae1 | Standard Test Method for Measuring the Curved Beam Strength of a Fiber-Reinforced Polymer-Matrix Composite |
| ASTM D 6641/D 6641M-09 | Standard Test Method for Compressive Properties of Polymer-Matrix Composite Materials Using a Combined Loading Compression (CLC) Test Fixture |
| **Some ASTM standards for composite flexural and shear properties** | |
| ASTM D 790-07e1 | Standard Test Methods for Flexural Properties of Unreinforced and Reinforced Plastics and Electrical Insulating Materials |
| ASTM D 2344/D 2344M-00(2006) | Standard Test Method for Short-Beam Strength of Polymer-Matrix Composite Materials and Their Laminates |
| ASTM D 3518/D 3518M-94(2007) | Standard Test Method for In-Plane Shear Response of Polymer-Matrix Composite Materials by Tensile Test of a ±45° Laminate |
| ASTM D 3846-08 | Standard Test Method for In-Plane Shear Strength of Reinforced Plastics |
| ASTM D 5379/D 5379M-05 | Standard Test Method for Shear Properties of Composite Materials by the V-Notched Beam Method |
| ASTM D 5448/D 5448M-93(2006) | Standard Test Method for In-Plane Shear Properties of Hoop-Wound Polymer-Matrix Composite Cylinders |

Source: Ref 8.10

usually significant in IR&D investigations rather than for quality control. ASTM D 2343 gives a test standard for glass-fiber strands that have (importantly) been impregnated with resin. There is a translation efficiency in proceeding from bare fiber to resin-impregnated fiber to composite laminate. Impregnating the fibers with resin before test provides a more realistic goal of potential fiber strength to be attained in the laminate and is a reason that ASTM D 3379, which addressed tensile testing of a single filament, has been withdrawn.

## Resin Properties

As mentioned previously, most value is derived in testing the fiber and the resin together, in the composite form. Many tests of neat resin specimens remain important, especially in developing processing methods. The ASTM standards D 638, D 695, and D 790 in Table 8.1 provide tests for the simple mechanical properties of tension, compression, and flexure of neat resin specimens.

## Test Specimen Configuration

Filament winding is most easily done over a mandrel with convex curvature, so that the tension of the fibers over the mandrel provides positive pressure to compact the laminate during winding, B-stage, and cure. Usually, fiber tension provides sufficient compaction so that no external pressure, as from an autoclave, is required during the cure process. This is an important advantage of filament winding, because very large parts, which would not fit into the largest autoclave, can be fabricated by filament winding. Unfortunately, filament-wound parts usually have higher void content and lower through-the-thickness properties than autoclaved parts.

Laminate void and fiber content have important effects on properties that are significantly influenced by the resin system, such as transverse shear strength, in-plane compressive strength, transverse tensile strength, and so on. One technique for measuring composite void content is ASTM D 2734, but this technique is applicable only where the effect of ignition on the materials is known, as with glass fibers. Micrographic techniques or image analysis (Ref 8.7, 8.12) use highly enlarged micrographs to determine void and fiber content.

Chemical matrix digestion assumes zero void content and determines fiber volume content by matrix digestion. The chemical reactions involved must be known for this technique to be useful, because the digestion of fiber by itself is liable to be different than in the laminate. Hot nitric acid has been found adequate for graphite/epoxy composites (Ref 8.7).

Figure 8.1 (Ref 8.13) shows that interlaminar shear strength drops steeply as void content increases for graphite/epoxy unidirectional laminates. Later studies on the same material, AS-4/3501 laminate, by Ghiorse (Ref 8.14) (Fig. 8.2) showed a drop of 9.7% void for interlaminar shear and a 10.3% void for flexural strength. Ghiorse also compared several density-measuring and two image analysis techniques. As discussed previously, filament-wound parts usually have higher void content than press- or autoclave-cured laminates. For example, the typical void content of high-void areas of an as-wound filament-wound rocket motor case can be 5 and perhaps 10%; pressure-cured laminates are rejected if the void content exceeds 1 to 2%.

This difference in acceptance criteria reflects the end use of the parts. In applications where fiber-dominated properties such as in-plane tensile strength are most important, as in rocket motor cases, higher void contents are accept-

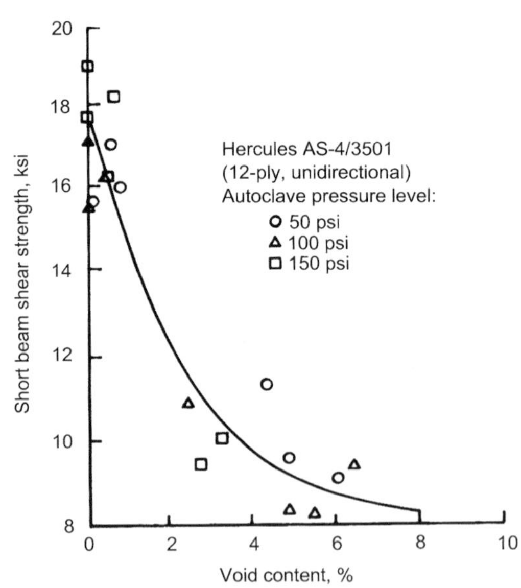


**Fig. 8.1** Effect of void content on shear strength. Source: Ref 8.13

able, and extra laminate compaction is not necessary. One application survives with virtually 100% voids. An isotensoid liquefied petroleum gas pressure vessel (Ref 8.15) is dry wound (no resin) over a thermoplastic liner and goes through a series of product tests. The operating pressure is 8 to 10 bar (116 to 145 psi), and the wound vessel has some distinct advantages over the steel version in terms of mass, fail-safe operation, impact resistance, and fire and flame resistance. Because the vessel is dry wound, winding speed can be drastically increased, and the vessel, after impact with a wedge at 29 mph (46.7 km/h), will hold pressure and return to original configuration under pressurization with no permanent deformation.

In applications where resin-dominated properties such as transverse shear strength are important, laminate compaction devices may have to be used. Procedures have been used to incorporate autoclave cure in the fabrication of filament-wound components, thus providing lower void content in filament-wound laminates when this is required. However, care must be exercised to avoid crimping the fibers in the composite.

A test specimen configuration must be selected on the basis that test results provide material properties useful in design. The specimen material must reflect the as-processed makeup of the material, so that the resulting test data can be used with confidence. Figure 8.3 shows a schematic of (I) flat panel, (II) ring, and (III) tubular specimens and how they may be tested. Some aspects of specimen selection procedures are described as follows.

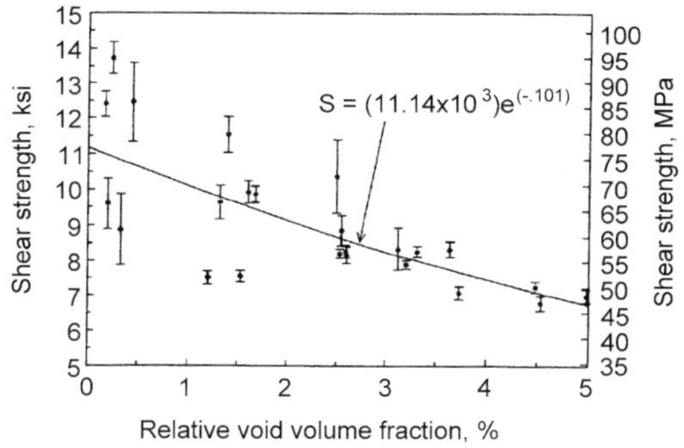


**Fig. 8.2** Effect of relative void content on shear strength. Source: Ref 8.14

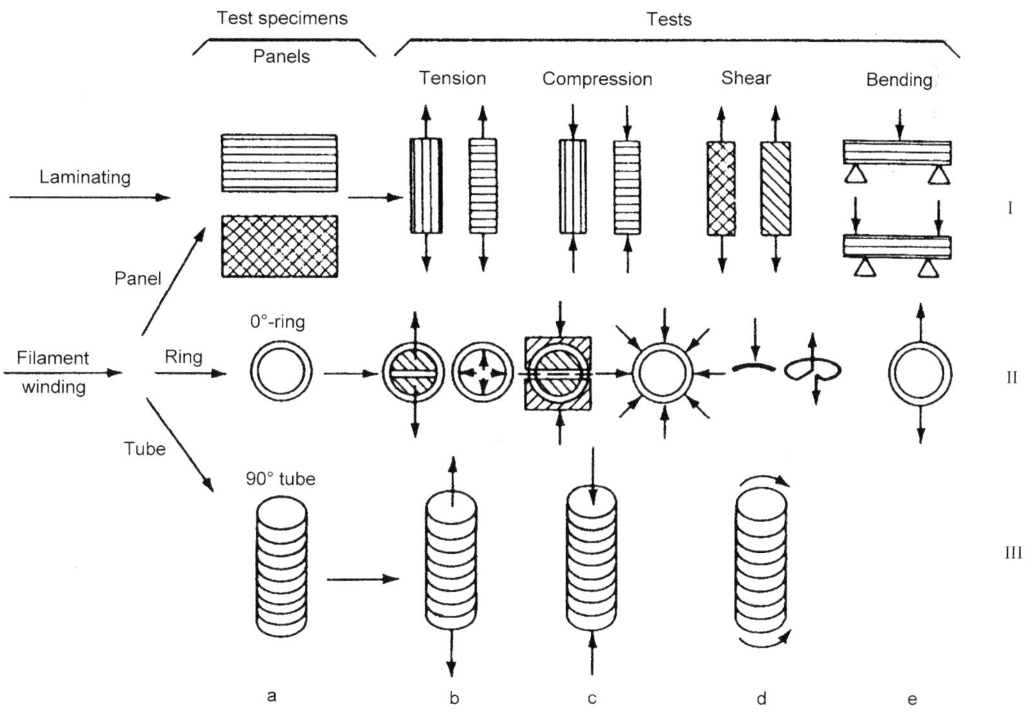


**Fig. 8.3** Test specimens and test techniques for filament-wound composites. Source: Ref 8.16

### *Flat Specimens*

Group I in Fig. 8.3 shows testing configurations for mechanical testing of flat panel specimens (Ref 8.16). Filament wound structures, if they are geometrically symmetric, as are pressure vessels, do not need a symmetric laminate layup because the structure satisfies that need. In fact, an unsymmetric laminate may confer some manufacturing advantages (i.e., putting hoop layers on the exterior will increase compaction pressure and may lower resin fraction, or putting helical layers on the outside of a pressure vessel will prevent impact damage to the usually higher-stressed hoops). Thus, duplicating the laminate for testing purposes will most probably require changing of the laminate layup, because the flat laminate will have to be symmetric. If a sample of the pressure vessel is to be cut out for test specimens, it will invariably warp unless the laminate is thick enough to resist the warping forces.

How can the filament-wound laminate be replicated for testing? One process for doing this involves winding impregnated unidirectional tow or roving over a large mandrel; cutting and removing the fibers, as-wound material, from the mandrel; and then plying, consolidating, and curing (in an autoclave or press) to the flat configuration. The layup of the panel may have to be changed to achieve symmetry, and the helical angle layers, for carbon-graphite laminates, should be wound as polar plies. The two helical plies in a layer are balanced at the same angle ± relative to the mandrel centerline and density but are not symmetrical because they cannot be wound with the complementing ply reversed. A thin, flat laminate for testing that contains carbon helical lamina may warp.

The material, as a result of this process, could be quite different from the material in a filament-wound composite structure. Using flat panels to characterize filament-wound parts creates several problems in addition to those cited previously, including:

- *Edge effects:* Interlaminar stresses, which can precipitate premature failure, arise at the free edge of a laminated composite. This is due to the different behavior tendencies of the multiangle plies interacting in the laminate (Ref 8.17).
- *Fiber wash:* Once cut, the fibers are not tensioned and thus are free to move, relocate, and lose straightness.
- *Void/fiber content:* The filament-wound laminate prepared as a flat panel may have different void and fiber contents than the original as-wound material.

Statistical aspects of strength and elastic properties were determined for T-400/LY 564 laminates with ±55 and ±45° wind angles for tensile and compression of flat panel test specimens. Philippidis et al. (Ref 8.18) then used six statistical distributions to model the mechanical properties. Because of the scatter, the coefficient of variation is greater than the mean value; in some cases, the models were all judged to be relatively equivalent.

Unidirectional prepreg and prepreg tow of T-650-35/1902 were fabricated by Jensen et al. (Ref 8.19) as several angle plies, both as hand-laidup flat specimens and as tubular filament-wound preforms that were then cut from the mandrel and hand laidup. Both sources of preform were then autoclave cured in accordance with the manufacturer's directions to result in flat panels. These flat panels were compression tested using the National Aeronautics and Space Administration (NASA) short-block fixture to identify the failure mechanisms and the strength and stiffness of large-scale fiber undulations. There were several ply orientations and stacking sequences. Jensen et al. found that the compressive properties and failure modes of the filament-wound specimens were degraded in the same manner as hand-laidup specimens, that the main effect of the undulations was to initiate failure, that there were several failure mechanisms, and that the adjacent layers to the crossovers had a strong influence on compressive strength, stiffness, and failure mechanisms by weakening or reinforcing the laminate.

### *Ring Specimens*

Techniques for fabricating ring-type test specimens are outlined in ASTM D 2291-03. The standard ring, commonly referred to as a Naval Ordnance Laboratory ring, is 5.75 in. (146 mm) inside diameter, 0.9 in. (22.8 mm) wide, and 0.06 in. (1.52 mm) thick. New ring specimens have been added to the ASTM D 2290-04 specification (Fig. 8.4). There are three test specimen configurations/procedures. Procedure A is for reinforced thermoset pipe, and B and C are for unreinforced thermoplastic pipe and have different dimensions. Procedure A has an option of one or two reduced sections that are to be

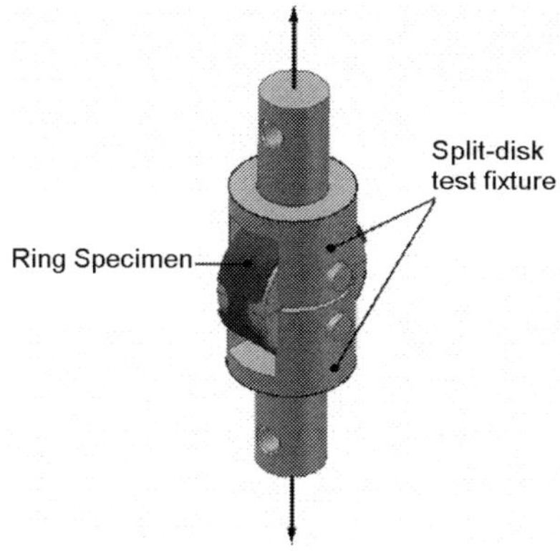


**Fig. 8.4** Split ring in accordance with ASTM D 2290-04. Source: Ref 8.20

mounted for testing 2 in. (50.8 mm) away from the split in the lubricated disk. Group II in Fig. 8.3 shows mechanical testing configurations for ring specimens.

### *Tubular Specimens*

The thin-walled tube is a desirable test specimen in that it provides a laminate that accurately represents filament-wound parts (group III in Fig. 8.3). To test unidirectional ply properties, the tube may be wound with hoop fibers only. In some cases, a laminated tube with interspersed [±θ] or [±θ/90] construction is used, as discussed subsequently.

When no standard exists, a stress analysis of the specimen and loading condition is required in order to understand the stress and strain distributions produced during test. There are some complications due to end fixity, and several papers in the literature (Ref 8.21, 8.22) provide guidance on specimen geometry. Some specimen configurations are provided as follows, along with suggestions concerning required geometry under specific loading conditions. It is important to note that laminate symmetry, essential for flat laminate testing, is not required for uniform strain in a tubular specimen. In the tubes, the bending-extension coupling induced by lack of symmetry occurs near the ends and rapidly diminishes in magnitude within the grip-influenced region, providing a test section with uniform strain.

Thin-walled tubes were filament wound, cured, and tested to result in smooth outer-wall test specimens for torsional shear tests. Dickson et al. (Ref 8.23) found that several techniques of curing the AS-4/3501 tubes resulted in fiber wrinkles or thin or thick longitudinal resin seams. Dickson et al. finally settled on wrapping 0.004 in. (0.1 mm) thick steel shim stock and then wrapped shrink tape with Kevlar (Du Pont) fiber dams, which produced acceptable surface finish and consistant shear strengths.

In 1984, Swanson et al. (Ref 8.24) tested thin-walled, long, hoop (1.14° from hoop direction) filament-wound tubes (4 in. outside diameter by 3 in. long and 0.1 in. wall, or 100 mm outside diameter by 75 mm long and 2.5 mm wall). Torsion tests involved several epoxy systems and two fibers, Kevlar 49 and AS4-12k, at a fiber volume fraction of 50%. The cured cylinders were bonded to aluminum end plates. Test results for shear stress versus shear strain closely followed curve-fit micromechanics predictions.

Spencer and Hull (Ref 8.25) tested E-glass/polyester tubes with four wind angles, ±35, 45, 65, and 75°, in modes 2 and 3 with an internal sealed elastomeric bag to allow test-through weepage to burst. They found that a strain normal to the fibers of 0.2% would induce weeping. The measured elastic properties were consistent with theory.

Wind angles of ±45, 55, 60, 75, and 88° were studied for glass-reinforced pipe made from E-glass and epoxy. Onder et al. (Ref 8.26) used the Lekhiniskii theory for analysis, along with finite-element methodology (FEM), to predict burst pressures of symmetric and nonsymmetric four-layer laminates. The tubular test specimens were closed end. They found that the optimum winding angle for internal pressure was ±55°, and that closed-form analysis gave a much better match to the test results than FEM, which used a first-ply failure criterion.

### *Flywheels*

A subscale high-speed rotor test specimen was used to characterize the performance of two epoxy tow-pregs and three fibers, T-650-42, IM7, and M46J. Nagy et al. (Ref 8.27) used a molybdenum-disulfide-filled nylon as the hub material, which then radially mass loaded the ring as well as the centrifugal forces due to the rotation of the thin composite ring. The results seemed to show that apparent fiber knockdowns should be on the order of 60%.

Gabrys and Bakis (Ref 8.28) developed a method of using a solvent-diluted polyurethane resin elastomer for rotors to result in a hoop-wound elastomeric-matrix composite to avoid the catastrophic failure associated with rigid epoxy-matrix composites. They were able to continuously wind hoop layers at accretion rates of over 1 in./h (3 cm/h) to high fiber volume (60%) and a void content of less than 1%. They wound with OC 463 750-yield S2 glass and AS4-GP 12k carbon. They took samples for rings to determine density, fiber volume, and void volume fractions as well as mechanical properties and coefficent of expansion.

Flat test specimens of IM7/8552 were wound and cured to approximate the configuration of the rotor and were tested in-plane with and transverse to the fibers to determine the time-temperature-dependent responses. Thesken et al. (Ref 8.29) tested compression and tensile properties at room temperature, 203 °F (95 °C), and 275 °F (135 °C) and found large transverse compressive stress relaxation (16.4%) at the higher temperature. Thesken et al. showed that the time-temperature superposition principle may apply for the temperature regimen selected.

Gabrys and Bakis (Ref 8.30) used AS4 and a bisphenol A epoxy resin to demonstrate high-speed winding with parallel curing. They achieved a radial deposition rate with hoops at 8 in./h (20.3 cm/h) with no apparent fiber buckling or waviness. They then measured density, fiber volume, and void fraction along with residual stresses, coefficient of thermal expansion, and transverse strength and stiffness.

There was a study by Takahashi et al. (Ref 8.31) that sought to fabricate a rotor which could achieve a peripheral speed above 4300 ft/s (1300 m/s). They tried several approaches to use several rings and two fibers, M40J and T1000G, and epoxy resin and several combinations for the multiringed rotors. They used a 7075 aluminum hub and machined the outside diameter of the outer ring. When they rewound the machined outer surface of the outer ring, they attained the target peripheral speed.

### *Other Structures—Drill Risers*

Subscale composite production drill risers of 10 in. (257 mm) nominal inside diameter using a multiple traplock design for the end fittings were tested for ultimate pressure with and without end-load effect by Salama et al. (Ref 8.32). Pressure at failure with a combination of internal pressure and axial load was below the test goal, and some redesign of the metal composite interface was necessary. After redesign, the riser met all performance and cost requirements.

### *Specimen Conditioning*

In most modern composites, especially those with polymeric matrices, the mechanical properties are influenced by the temperature and moisture content of the material. ASTM D 618 provides standard methods of conditioning composite materials to produce consistent results.

## Unidirectional Ply Properties

The general approach in mechanical testing is to characterize the unidirectional ply, the basic building block of the material. Then, through lamination theory, the behavior of the laminate can be predicted. The mechanical properties of the unidirectional ply required for characterization are:

- Four elastic constants (sometimes, compressive as well as tensile modulii are determined):
  1. Longitudinal modulus of elasticity, $E_1$
  2. Poisson's ratio, $\nu_{12}$
  3. Transverse modulus of elasticity, $E_2$
  4. In-plane shear modulus, $G_{12}$
- Four hygrothermal constants:
  1. Longitudinal coefficient of thermal expansion, $\alpha_1$
  2. Transverse coefficient, $\alpha_2$
  3. Longitudinal coefficient of moisture expansion, $\beta_1$
  4. Transverse coefficient of moisture expansion, $\beta_2$
- Five strengths:
  1. Longitudinal tensile strength, $F^{t}_{L}$
  2. Longitudinal compressive strength, $F^{c}_{L}$
  3. Transverse tensile and compressive strength, $F^{t}_{T}$
  4. Transverse compressive strength, $F^{c}_{T}$
  5. In-plane shear strength, $F^{s}_{LT}$

ASTM International standard test methods for characterizing composite materials are listed in Table 8.1. Some of these, as well as other methods that are not standards, are described subsequently when test methods are discussed.

### *Longitudinal Tension Properties, $E_1 \nu_{12} F^{t}_{L}$*

The preferred test for determining longitudinal tensile properties of the fiber itself is internal pressurization of a hoop-wound tube (Fig. 8.3, group III) of radius $R$ and wall thickness $t$. The test fixturing is such that the fixture, rather than the specimen, carries axial pressure loads; Fig. 8.5 shows such a device. For a thin-walled tube with uniaxial hoop stress due to internal pressure, $P$:

$$E_1^t = \frac{PR}{\varepsilon_y \cdot t} \quad \text{(Eq 8.1)}$$

where $\varepsilon_y$ is the measured hoop strain. Also:

$$\nu_{12} = \frac{\varepsilon_x}{\varepsilon_y} \quad \text{(Eq 8.2)}$$

where $\varepsilon_x$ is the measured axial strain. At burst pressure:

$$F_L^t = \frac{P_{max} \cdot R}{t} \quad \text{(Eq 8.3)}$$

The contribution of longitudinal reinforcement is negligible but does serve as a crack stopper to prevent premature failures.

It is important that the tube is thin, so that measured hoop strains at the outside surface are approximately equal to the (maximum) hoop strains at the inside. Some reinforcement at the ends, and possibly longitudinal reinforcement throughout the tube, can be added to control discontinuity effects at the ends and to restrict radial stress effects on failure. Tube dimensions are chosen so the central test section takes on a uniform strain state, free from effects due to the end plugs. Bending effects rapidly reduce to less than 1% of the peak value away from the ends for symmetrical laminates (Ref 8.33).

Conservative estimates (Ref 8.17, 8.22) indicate that this is accomplished with a radius-to-thickness ratio greater than 10 and a tube length greater than the desired gage length plus 4 times the radius, depending, of course, on end fixity induced by the test fixturing.

Another method of measuring fiber strength is the pressurized ring specimen (Fig. 8.3, group II-d), which is less expensive than the tube but allows no axial reinforcement or axial strain measurement. Interpretation of the test is the same as for the tube, using Eq 8.1 and 8.3. Ring pressurization testing is preferred to split-disk tension testing (Fig. 8.6 and ASTM D 2290), which can introduce high bending stresses at the split between the split-disk fixtures.

A pressurized 20 in. (50.8 cm) diameter ring test with an internal bladder for cleanup convenience was used by Cohen et al. (Ref 8.34) to bridge the gap in test results between single-filament, fiber strand, or coupon tests and full-scale tests. The 20 in. (50.8 cm) diameter size was based on a commonly used size for subscale pressure vessels. They found that the results compared well to the full-scale vessel, reducing the sizing effort as well as costs. The two ASTM International specifications that governed the fabrication and testing were ASTM D 2290 and ASTM D 2585 (now withdrawn).

Arsene and Bai (Ref 8.35) did not test composites with the ISO 8496-1986 ring test; they

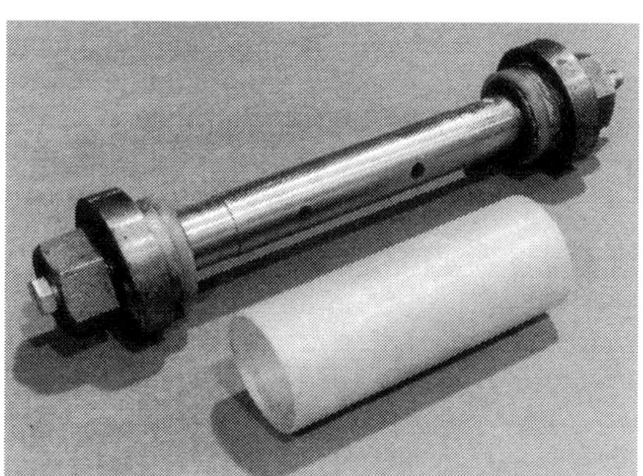

**Fig. 8.5** Test fixture for pressurized hoop-wound tube. Courtesy of A.O. Smith Co. Source: Ref 8.3

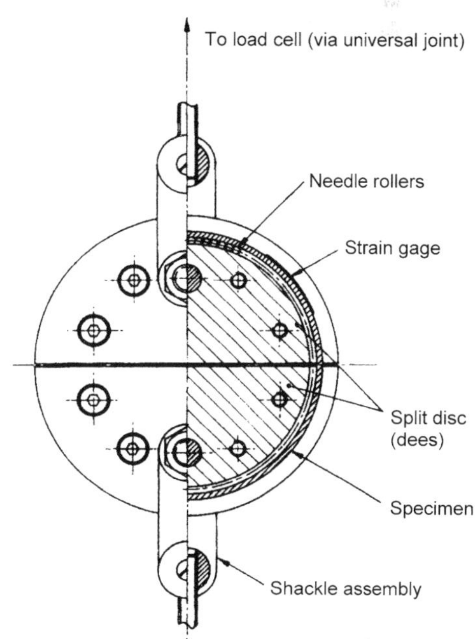


**Fig. 8.6** Roller-assisted split-disk apparatus. Source: Ref 8.20

tested the isotropic metal Zircaloy to optimize the test for composites and compared the results with finite-element analysis (FEA). The results of the effort were recommendations for changes in the test method, that is, modification of the ring geometry to achieve more uniform stress distribution and a stable, longer gage length in the test specimen.

Cohen (Ref 8.36) wet wound eight 20 in. (50.8 cm) diameter graphite/E-glass composite cylinders and then cut 1 in. (24.5 mm) wide ring test specimens. Twelve variables were evaluated; five variables were chosen for study in a design-of-experiment-controlled experiment. The five manufacturing variables studied were stacking sequence, cut versus uncut helicals, winding band tension, winding tension gradient, and winding time between layers. The tests were pressurized ring for hoop strength and stiffness, short-beam shear, dimensions, image analysis for fiber volume and void content, and split ring to evaluate residual stresses. Cohen found that composite strength was significantly affected by stacking sequence, winding tension, winding time, and the interaction between winding tension and winding time. High winding tension produced higher fiber volume, and shorter winding time resulted in higher fiber volume than long winding time.

External pressure can be applied to a composite ring by axial load to a compliant ring of high-density polyethylene that is sandwiched between the composite and a rigid steel outer ring. Kugler and Moon (Ref 8.37) tested T-300/Udel P1700 (Solvay Advanced Polymers) polysulfone thermoplastic composite at two hold temperatures, two tow tensions, and several *R*/*t* ratios. The technique was described as an alternate, lower-cost analysis technique compared to cylinder tests for both thin and thick composite rings.

Diametral compression testing of ring composites is covered by ISO 7685 (Ref 8.38). Part of the analysis of the ISO 7685 method was carried out by Pierron and Davies (Ref 8.39). They tested glass/epoxy rings at various thicknesses and at wind angles of ±0, 30, 45, 60, and 90° to determine material modulus, which was not the aim of the ISO test procedure. They found that for some layups, such as $[\pm\theta]_n$ or $[\pm\theta]_{ns}$, the bending modulus is not equal to the in-plane modulus, and the two tests should produce the same modulus value only if there is a correction factor that takes into account the external resin coating in the FEA.

The unidirectional carbon fiber (Toray T700) epoxy rings used for radial containment of a ultrahigh-molecular-weight polyethylene (UHMWPE)-bearing element had some circumferential splitting under full-scale load. Samyn et al. (Ref 8.40) determined short-beam shear and transverse compression strength of samples taken from rings with tests modified from the ASTM International standards because of available sample configuration. They did FEA on the subscale test coupons, full-scale UHMWPE, ring laminate, and steel retainer. None of the analyses revealed an unambiguous single cause of failure, but rework of the geometry, modification of the winding process to reduce the fiber volume from 69 to 66%, and reduction of the void content from 3.5 ± 3% to 1.5% allowed full-scale tests without carbon/epoxy failure.

Yoon et al. (Ref 8.41) were able to measure the tensile strength and modulus of ring specimens fabricated from an epoxy resin system and T-300, T-800, T-1000, and S2 glass fibers. They modified the ASTM D 2290 fixture to eliminate bending. They used eight strain gages around the periphery of the straight-sided test specimens and averaged the modulii from the loading and unloading of rings at the same crosshead speed.

A study by Cohen (Ref 8.42) sought to statistically relate the winding tension, stacking sequence, winding-tension gradient, winding time, and the interaction between winding-tension gradient and winding time. Eight 20 in. (50.8 cm) diameter by 30 in. (76.2 cm) long cylinders were wound with two helical angles as well as hoops and then cut into 1 in. (2.54 cm) wide rings that were evaluated via a pressurized ring test, short-beam shear, and image analysis to determine fiber volume and void content. Winding tension, laminate stacking sequence, winding-tension gradient, and winding time were found to significantly affect pressure vessel strength. Higher winding tension and shorter winding time were found to result in higher fiber volume. Residual strains and through-thickness material properties, radial Young's modulus, and coefficient of thermal expansion (CTE) were measured for thick wound rings made from E-glass and epoxy wet wound at 90, ±75, and ±60° wind angles. Young's modulii varied slightly, but there was a large increase in the radial CTE and a large decrease in the radial strength over the wind angles. Using a split ring, Ha and Jeong (Ref 8.43) also found a significant decrease in residual strains as the wind angle decreased.

Some investigators have suggested alternate means of carrying out the split-disk technique to obtain better mechanical values. The roller-assisted variant of the split-disk test was investigated by Jones et al. (Ref 8.20) to attempt to equalize the strain throughout the wall of the ring. They included a set of needle rollers between the disk and the ring test specimen (Fig. 8.6). They modeled the ring test, performed an FEA, tested rings with wind angles of ±45, ±60, and ±75°, and compared the results to those from closed-end cylinders of the same material and wind angles. The cylinder and ring results were consistant for the ±75° wind angle but did not track for the ±45° wind angle.

Kaynak et al. (Ref 8.44) tested five different fibers, two epoxy systems, and five wind angles. They aligned the reduced section of the ring test specimens with the split of the fixture, and they measured hoop tensile strength and modulus but did not record fiber volume. It is well known that fiber wind angle influences the fiber volume and that a higher wind angle will result in higher fiber volume for a wet-wound composite, all other factors remaining constant.

Tse et al. (Ref 8.45) tested nominal 4 in. (100 mm) diameter by 2 in. (50 mm) long by 0.1 in. (2.5 mm) thick glass epoxy rings with wind angles of ±45 and ±75° and five symmetric layers (covers) to determine in-plane stiffness ($K_y$), in-plane bending-shear stiffness ($K_{xy}$), and bending-torsional stiffness ($K_{xz}$). They found good agreement between the closed-form and FEA and between the experimental results.

Strand tensile tests (e.g., ASTM D 4018) and flat coupons (Fig. 8.3, group I-a and ASTM D 3039) provide tensile property data. Strand tensile tests produce higher-than-operational allowables, and, as discussed previously, the flat coupon is probably not typical of as-wound filament-wound laminates.

### *Transverse Tension Properties, $-E_2\nu_{21}F^t{}_T$*

The preferred test for determining transverse tensile properties is axial tension of a hoop-wound tube (Fig. 8.3, group III-b). Figure 8.7 shows the configuration for a transverse tension tube in accordance with ASTM D 5450. The potted-in ends are designed to provide sufficient strength so that failure occurs in the test section. Usually, as in Fig. 8.7, the ends are not reinforced.

This test specimen will also suffice for two other tests of interest: transverse compression in accordance with ASTM D 5449 and in-plane shear in accordance with ASTM D 5448. Each of these standards provides recommendations for test analysis and a description of failure modes using common test apparatuses and test specimens.

Another test for transverse strength of a composite is ASTM D 6415. The specimen is a 90° bend with a 0.25 in. (6.4 mm) inside radius and 3.5 in. (90 mm) long legs (Fig. 8.8). This test is accepted in the composite community but is difficult to build by filament winding. The filament-wound fiber, under winding tension, tends to have more compression over the tight radius, resulting, if not accounted for, in lower resin content in the bend. Some improvement can be obtained by restricting the resin flow in the legs by vacuum bag or shrink tape during cure.

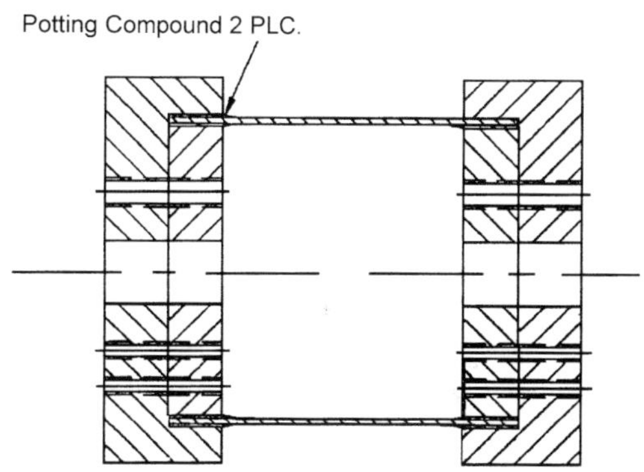


**Fig. 8.7** Tubular test specimen in accordance with ASTM D 5450. Source: Ref 8.3

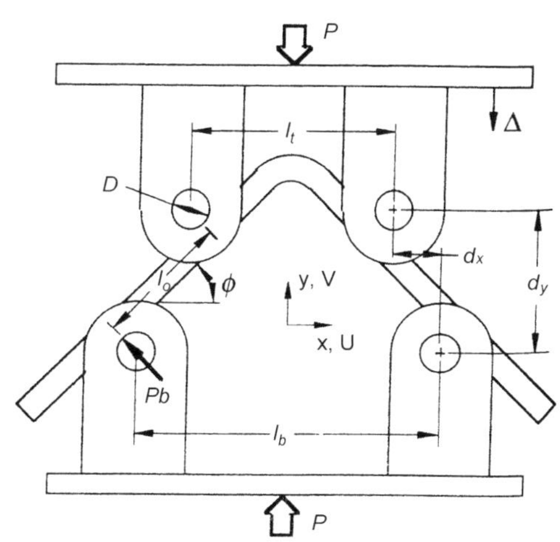


**Fig. 8.8** ASTM D 6415 curved-beam test specimen for interlaminar tensile strength. Source: Ref 8.10

The variant of this test, more amenable to filament winding, is the horseshoe-shaped curved beam, also with straight sides (Fig. 8.9, Ref 8.46), where $\sigma_r$ is the radial stress, $P$ is the applied end force, $b$ is the outer radius of the semicircular region of the horseshoe-curved bar, and $a$ is the inner radius of the semicircular region of the horseshoe-curved bar. Ko and Jackson used Lekhnitskii's anisotropic stress functions and Ko's multilayer theory, along with FEA and testing, to verify the approach. Testing was done on samples with 22 unidirectional circumferentially oriented plies with a thickness of 0.125 in. (0.3175 cm). They found that the open-mode delamination stress could be greatly increased by decreasing the aspect ratio, $b/c$. Although a simple test, by virtue of the relatively easy fabrication process and loading mechanism (simple tension on the legs), the test has been reported to result in some significant nonuniform stresses (Ref 8.47).

Several tests have been introduced to measure the transverse tensile strength of a curved beam. Thick-walled laminates have become increasingly important in configurations such as driveshafts and flywheels. The common failure mode for rotating shafts and flywheels is radial tensile stress, which peaks near the midthickness of the rotating body. ASTM D 7291 test method was developed to address the problems of the curved-beam transverse tests. However, the test specimens have some potential handling problems, and ASTM International avers that the values would not correspond to those of the curved-beam test.

The final test, proposed by Sharma and Bakis (Ref 8.48), also uses a C-shaped test specimen. They used FEA to determine the residual stresses and to predict the effects of eccentric loading. Hoops of T700/RF007 and other epoxy resins and fibers were wet wound over a cylinder, cured, and then cut into individual rings. A circumferential section was machined on each ring to produce the C-shape with parallel edges. Residual stresses were evaluated by sectioning rings that had three foil strain gages on their outside diameter. The test specimens were loaded with a hydraulic jack with crossed steel rollers on one side of the jack to eliminate moments. Negligible residual stresses were found. Finite-element analysis and experiments showed that the test is tolerant of eccentricities. The transverse strengths of the rings were generally less than the strengths measured by flexure specimens.

### *Longitudinal Compression Properties,* $-E_1\nu_{12}F^{c}_{L}$

Test techniques for determining compressive design allowables are still a topic actively being discussed. Standard test methods (e.g., ASTM D 3410) provide a specimen configuration with a very short unsupported length to avoid delamination, local buckling, Euler buckling, and other types of failures. Unfortunately, this suppresses

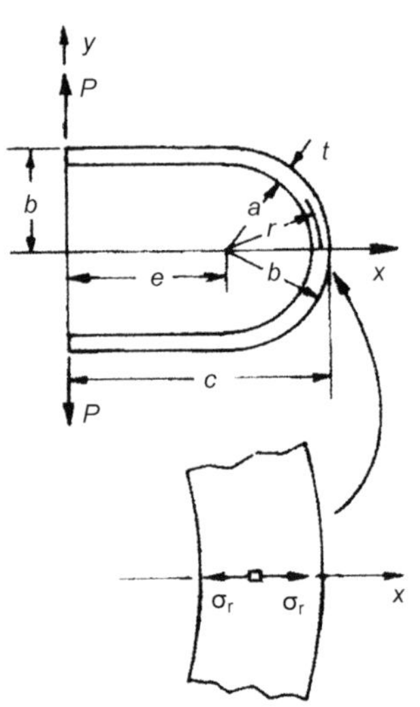


Analysis of test results for this specimen involves the relatively simple equation:

$$\sigma_r = \left[ \frac{P}{h(b-a)} \right]$$

Where:

$\sigma_r$ = radial stress
*scf* = stress-concentration factor
$P$ = applied end force
$h$ = width of semicircular bar
$b$ = outer radius of semicircular region of horseshoe-curved bar
$a$ = inner radius of semicircular region of horseshoe-curved bar

The *scf* is a function of the material being tested, the aspect ratio, and several other factors.

**Fig. 8.9** Horseshoe-curved beam for interlaminar tensile strength. Source: Ref 8.46

many of the failure mechanisms that occur in actual designs. Reported fiber strain values from ASTM D 3410 tests (≈1.2% for HS graphite epoxy) are much higher than allowables attained in actual parts (≈0.8%). As a result, many design engineers are reluctant to use compressive strength design allowables resulting from these standard techniques. Instead of using ASTM D 3410, some investigators use a modification of this specification or a modification of ASTM D 695, which is not intended for highly oriented fibrous composites. This is an example of the system at work. The standards are a consensus of opinion; if a test technique is widely used for valid reasons, it may eventually become a standard.

A test for determining longitudinal compressive properties for filament-wound laminate is the laminated tube of interspersed helical hoop construction loaded in axial compression. A tube of $[90_2/\pm20/90_{ij}/\pm20/90_2]_T$ construction, with a 2.8 in. (70.6 mm) inside diameter and a 0.1 in. (2.54 mm) wall thickness, has been used (Ref 8.49) and is typical of rocket motor skirt applications. The helical fibers, aligned at a low wind angle (e.g., ≈±20°) to the tube axis, carry most of the axial load. The hoop fibers, aligned at ≈90° to the tube axis, serve to stabilize the tube. The hoops are important. Tubes with only helicals and no hoops can fail, with the tube expanding in diameter with no fiber failure.

The NASA short-beam compression test can be used for either in-plane or transverse testing (Fig. 8.10), if a test specimen can be machined for use (Ref 8.50).

A ring specimen under external pressure (Fig. 8.3, group II-c) provides a viable specimen for determining longitudinal compressive properties of the unidirectional ply. Interpretation of the test is the same as for the pressurized ring and tube, using Eq 8.1 and 8.3, determining compressive rather than tensile properties. The ring pressurization testing is preferred to the split-disk compression testing (Fig. 8.3, group II-d) because of discontinuities introduced at the split between the compression fixtures.

**Fig. 8.10** NASA short-beam compression test fixture. Source: Ref 8.50

An ASTM International standard (ASTM D 2586) provides test methods for determining hydrostatic compressive strength of tubes in external pressure. In the standard, a tube with end plugs is placed in a pressure chamber, producing in-plane biaxial compression on the laminate and an additional radial compressive stress at the outside surface. Under these conditions, the method provides, as stated in the specification, "figure-of-merit" values and not true unidirectional properties.

### *Transverse Compression Properties, $-E_2\nu_{21}F^c_T$*

Axial compression of a hoop-wound tube in accordance with ASTM D 5449 (Fig. 8.3, group III-c) is the method recommended for determining transverse compressive properties. The specimen is the same as used for transverse tension (Fig. 8.7). There is no other preferred specimen geometry. The flat transverse specimen (Fig. 8.3, group I) is fragile and more susceptible to alignment difficulties than is the tube.

Transverse compression of thin-walled unstiffened tubes was studied by Messager et al. (Ref 8.51) for optimizing laminate angles and ply stackup. They used a generic algorithm, an analytical model of shell buckling, and FEM as well as testing to verify the analyses. They compared several optimized laminates to the standard $[\pm55]_d$, and the studies led to several optimized laminates: $[90_2/60/30_5/60/90]$ for T700 carbon fiber and $[90_3/15_2/90_2]$ for E-glass epoxy. A test setup was used that employed a standard tensile machine to supply the external hydrostatic pressure to the test specimen (Fig. 8.11).

### *In-Plane Shear Properties, $G_{12}$ and $F^c_{LT}$*

The preferred test for determining longitudinal shear properties of filament-wound composites is axial torsion of a hoop-wound tube in accordance with ASTM D 5448 (Fig. 8.3, group III-d). The specimen shown in Fig. 8.7 with potted-in ends has proved to be a satisfactory

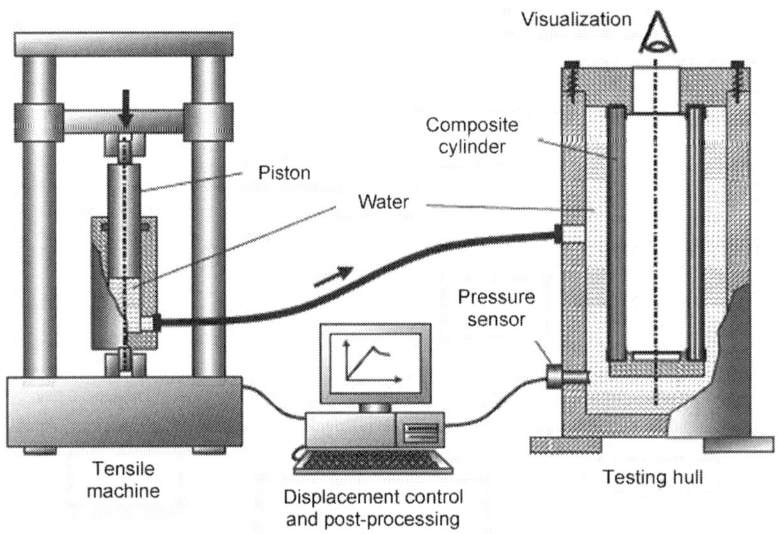


**Fig. 8.11** External compression device. Source: Ref 8.51

configuration. The details are found in ASTM D 5448.

ASTM D 3518 presents a standard practice for in-plane shear response of fiber-reinforced composites. The method is based on the response of a 45° symmetrical laminate loaded in uniaxial tension. This test can be extended to a thin filament-wound tube of ±45° construction loaded in axial tension (Fig. 8.3, group I-d).

The ring out-of-plane loading test (Fig. 8.3, group II-d) can be used to determine the longitudinal shear modulus of a unidirectional laminate. The short-beam shear test (Fig. 8.3, group II-d and ASTM D 2344) is primarily useful as a resin and processing screening device. The test is inappropriate for determining interlaminar shear strength because the stress condition at failure is too complicated to provide useful strength values.

The Iosipescu test in accordance with ASTM D 5379 is one of the accepted techniques for determining shear properties of composite materials. Originally developed for metals, the technique has been modified for application to composites, as summarized in Ref 8.52. The test is most effective with quasi-isotropic laminates and least effective with unidirectional laminates. Essentials of the technique are shown in Fig. 8.12. The bar specimen is double notched at the center. On application of the compressive forces, *P*, the notched section is subjected to a vertical shearing force, with no bending moment. Finite-element analysis has shown that close-to-uniform shear stresses are produced in the notched section.

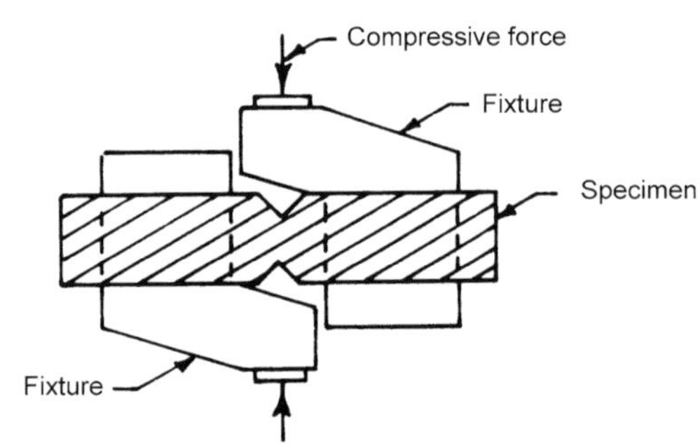


**Fig. 8.12** Schematic representation of Iosipescu shear test fixture and specimen (ASTM D 5379). Source: Ref 8.3

Some of the history of this test specimen development follows. An early critic of the short-beam test method was Keith Kedward (Ref 8.53), who suggested that the beam width should not exceed the depth for unidirectional carbon-fiber-reinforced plastic or glass-fiber-reinforced polymer and had reservations about the use of angle-ply laminates in this test.

The Iosipescu shear test has been championed in the United States since 1982 (Ref 8.52) by Adams as a simple and accurate substitute for the short-beam test. Along with Kedward, he has consistently critiqued the short-beam shear test (ASTM D 2344), even though it ranks the highest in usage due to its simplicity, lack of special test fixtures, and low cost. Adams has also suggested changes in the short-beam shear test procedure, that is, to use a span-to-thickness ratio of 6 to 8 rather than the ASTM International suggested ratio of 4 to 8, and to increase the diameter of the loading bar from 0.25 to 1 in. (6.35 to 25.4 mm) (Ref 8.54, 8.55). This critique of the short-beam test is well deserved. The Iosipescu test will give results for shear stress and stiffness, while the short-beam test is only useful for material screening, quality-control testing, and assessing the relationship between the fiber and the matrix in a unidirectional laminate. Both tests are simple, using an uncomplicated test fixture. The data from the ASTM D 2290 test can most appropriately be used to assess the ability of a resin to adhere to differing fiber surface treatments, but the test does not produce shear data for engineering design.

## Standard Pressure Bottles

Because of the inability to study dome contours, dome reinforcements, or skirts with 4 and 5.75 in. (101.6 and 146 mm) subscale bottles, such as the ASTM D 2585 design, the Air Force funded an effort to replace them with a test vessel that would more faithfully model actual flight hardware.

The single subscale test bottle to supplant the numerous rocket motor supplier designs was called the standard test and evaluation bottle (STEB) (Ref 8.56). The STEB (Fig. 8.13) is designed to serve as standard in the composite rocket motor case industry for comparing materials, process, and design features. The STEB is modeled as a one-ninth scale of a typical second-stage ballistic rocket motor case. The baseline configuration used Kevlar 49 fiber and an anhydride-epoxy matrix, with a 75% helical-to-hoop-fiber stress ratio, as determined by netting analysis. The cylinder length was selected to allow either polar or helical winding. Table 8.2 shows the wind parameters for both configurations. Alternate configurations, including different materials, processes, and design features, are easily evaluated using the STEB.

The STEB was developed and reported by Newhouse and Humphrey (Ref 8.56). It can be wound with skirts, can be helically or polar wound, and can support studies of the dome

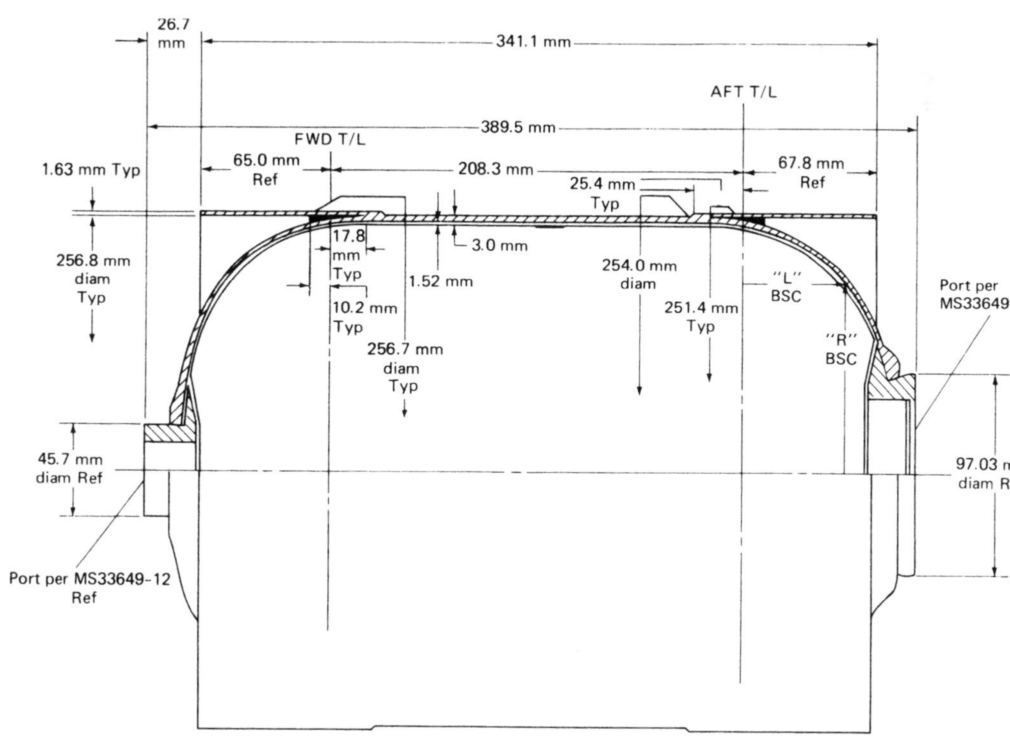


**Fig. 8.13** Standard test and evaluation bottle pressure vessel. Source: Ref 8.56

region and the dome/boss area as well as the cylinder. Further studies supported a filament-wound insulator study, investigation of premature forward boss blowout, stress ratio studies, among others, and several studies by outside agencies (Ref 8.57).

Zylon HM (Toyobo Co.) fiber and epoxy was hoop wound over an annealed copper open-ended cylinder that served as a mandrel and as a pressure-distributing agent when the cylinder ends were capped with a steel cylinder that restricted axial deformation. The high wind tension of 102 ksi (0.7 GPa) resulted in a fiber volume of approximately 90% and a developed stress for the composite of 696 ksi (4.8 GPa). Huang et al. (Ref 8.58) also discussed the beneficial effects of the prestress on the reinforcement.

The volumetric size effect of strength degradation with increased stressed volume was shown to be 16% from strand to unidirectional laminate and 32% to hoop laminate in a pressure vessel. Hwang et al. (Ref 8.59) tested impregnated tows, unidirectional laminates, and subscale STEB pressure vessels, all wet wound with one lot of fiber, T-800, and Epon 826 resin and the same cure cycle. They compared the Weibell weakest-link model and the sequential multistep failure model. They also averred that the fiber volumetric size effect is not a single phenomenon but depends on the material and processing factors.

Cohen et al. (Ref 8.60) employed a model and image analysis to predict and measure fiber volume fraction and to relate that to ultimate hoop strain. They used controlled fiber volume strand tests, subscale composite 20 in. (51 cm) rings with an $r/t$ ratio of 10, and data from full-scale tests to arrive at their conclusions. They found that when winding time is long, the particular resin used (HBRF 55) will start to gel and may result in a low-volume-fraction layer. The data showed that a 10% (or greater) improvement in composite fiber strength may be obtained by increasing fiber volume from 50 to 65%.

Park et al. (Ref 8.61) calculated the wind angle differences through the pressure vessel wall as a function of helical wind angle and ply thickness. They introduced the advanced standard test and evaluation bottle, a variant of the STEB, with a longer cylindrical section and different dome configurations, and compared those vessels, as shown in Table 8.3. They found as much as 18° wind angle difference between the first and last plies near the polar openings (Fig. 8.14). Pressure test results showed good agreement with the FEA. The use of dome caps and wafers was also investigated.

Basic design for a STEB was netting, and FEA was used to analyze two types of local reinforcement of the dome and two stress ratios, 60 and 88%. Hwang et al. (Ref 8.62) burst tested 13 subscale vessels that had been wet wound with T-800/epoxy. They concluded that a spiral-

**Table 8.2 Baseline standard test and evaluation bottle design (Kevlar 49, Du Pont)**

| Mechanical property | Helical | Polar |
|---|---|---|
| **Helical/polar fibers** | | |
| Design strength, ksi (MPa) | 320 (2200) | 320 (2200) |
| Wind angle, degrees | 18.0 | 12.5 |
| Fiber thickness, in. (mm) | 0.033 (0.828) | 0.034 (0.853) |
| Circuits to close pattern | 1.1 | 1 |
| Circuits per layer | 76 | 76 |
| Band width(a), in. (mm) | 0.40 (10.1) | 0.39 (9.9) |
| Rovings per band | 4 | 4 |
| Band density(a), ends/in./ply (ends/mm/ply) | 39.9 (1.57) | 41.1 (1.62) |
| Number of layers | 3 | 3 |
| Number of dome caps | 1 | 1 |
| **Hoop fibers** | | |
| Design strength, ksi (MPa) | 425 (2950) | 425 (2950) |
| Fiber thickness, in. (mm) | 0.04 (1.10) | 0.05 (1.20) |
| Band width, in. (mm) | 0.23 (5.8) | 0.21 (5.3) |
| Rovings per band(a) | 2 | 2 |
| Band density(a), ends/in./ply (ends/mm/ply) | 35.4 (1.39) | 38.5 (1.52) |
| Number of plies | 9 | 9 |
| Design burst pressure, ksi (MPa) | 4.2 (29.2) | 4.0 (27.6) |

(a) 12k roving, 6540 denier. Source: Ref 8.3

**Table 8.3 Comparison of standard test and evaluation bottle (STEB) and advanced standard test and evaluation bottle (ASTEB) subscale test pressure vessels**

| Property | Forward dome | | Aft dome | |
|---|---|---|---|---|
| | in. | mm | in. | mm |
| **STEB** | | | | |
| Cylinder radius, $R_c$ | 5.00 | 127.00 | 5.00 | 127.00 |
| Boss radius, $R_b$ | 0.8 | 22.05 | 1.9 | 47.70 |
| Cylinder thickness, $t_c$ | 0.06 | 1.4 | 0.06 | 1.4 |
| Hoop winding thickness, $t_h$ | 0.06 | 1.6 | 0.06 | 1.6 |
| **ASTEB** | | | | |
| Cylinder radius, $R_c$ | 5.00 | 127.0 | 5.00 | 127.0 |
| Boss radius, $R_b$ | 1.8 | 45.80 | 2.8 | 70.83 |
| Cylinder thickness, $t_c$ | 0.04 | 1.0 | 0.04 | 1.0 |
| Hoop winding thickness, $t_h$ | 0.044 | 1.113 | 0.044 | 1.113 |

Source: Ref 8.61

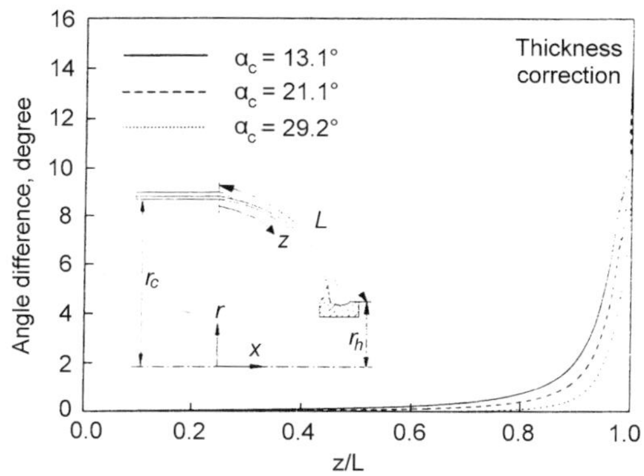


**Fig. 8.14** Winding angle differences between first and last plies with various winding angles. Source: Ref 8.61

wound wafer and a dome cap and a high stress ratio would be the most efficient design.

Finite-element analysis was used to predict the burst characteristics of a third-stage rocket motor case. The case, 20.5 by 20.3 in. (520 by 515 mm) with unequal dome openings, was wound with T-800 and an epoxy novolac resin. The analysis, by Park et al. (Ref 8.63), showed excessive deformation of the domes. The domes were reinforced, reducing the stress concentrations. Analysis predicted burst pressure of 4.7 ksi (32.64 MPa), and experimental results were 3 to 4.5 ksi (20.69 to 31.03 MPa).

Three pressure vessels with unequal polar openings were wet filament wound with T-800 carbon fiber/Epon 9405 epoxy resin. Doh and Hong (Ref 8.64) used degenerated finite shell elements and a progressive failure analysis for analysis. The cylinder radius was the same as a STEB, 10 in. (254 mm), but the vessel size and boss radii were greater. Hydrostatic tests resulted in dome-edge failure and an average burst pressure of 2 ksi (14 MPa).

### *Combined Load Testing of Tubular Specimens*

In addition to pressure bottles, tubular specimens have been developed as standard test specimens for combined load testing of composites. With tubes, axial load, in addition to internal pressure, can be applied conveniently. This allows characterization of the materials in the same combined stress environment as pressure vessels and rocket motor cases. Up until the year 2000, there was an unfilled need for standard test specimens and methods for testing filament-wound composites under combined loading, as in pressure vessel or rocket motor case applications. Although there is no standard, the World-Wide Failure Exercise (Ref 8.65) has filled that need, in part. Some of the test methods under development are described as follows.

Filament-wound tubes were a main source of test data for the World-Wide Failure Exercise, where 15 groups of researchers were tasked to predict deformation and failure of laminates in complex stress states using their own failure theories. Soden et al. (Ref 8.65–8.68), who organized the effort, then carried out the testing to validate the 19 submitted failure theories. Testing included uniaxial and biaxial tension, biaxial with compressive load, axial compression, and open-ended pressure, using a variety of test arrangements for both carbon and glass laminates (Fig. 8.15).

## REFERENCES


8.1 S.T. Peters, in *Reference Book for Composites Technology*, Vol 2, S.M. Lee, Ed., Technomic Publishing, 1989, p 143–178

8.2 R.F. Foral, in *Filament Winding: Composite Structure Fabrication*, S.T. Peters, W.D. Humphrey, and R.F. Foral, SAMPE Publishers, Covina, CA, 1994, p 6-1 to 6-24, © S.T. Peters, 2009

8.3 S.T. Peters, in *Filament Winding: Composite Structure Fabrication*, 2nd ed., S.T. Peters, W.D. Humphrey, and R.F. Foral, SAMPE Publishers, Covina, CA, 1994, p 7-1 to 7-31, © S.T. Peters, 2009

8.4 S.T. Peters and W.D. Humphrey, in *Composites*, Vol 1, *Engineering Materials Handbook*, T.J. Reinhart, Ed., ASM International, 1987, p 501–517

8.5 S.T. Peters and J.L. McLarty, in *Composites*, Vol 21, *ASM Handbook*, ASM International, 2001, p 536–549

8.6 J.M. Whitney, I.M. Daniel, and R.B. Pipes, "Experimental Mechanics of Fiber Reinforced Composite Materials," SESA Monograph No. 4, The Society for Experimental Stress Analysis, Brookfield Center, CT, 1982

8.7 L.A. Carlsson and R.B. Pipes, *Experimental Characterization of Advanced Composite Materials*, Prentice-Hall, Englewood Cliffs, NJ, 1987

8.8 *Military Handbook 17*, C Revision (MIL-HDBK-17C), Technomic Publishing and Materials Sciences Corporation

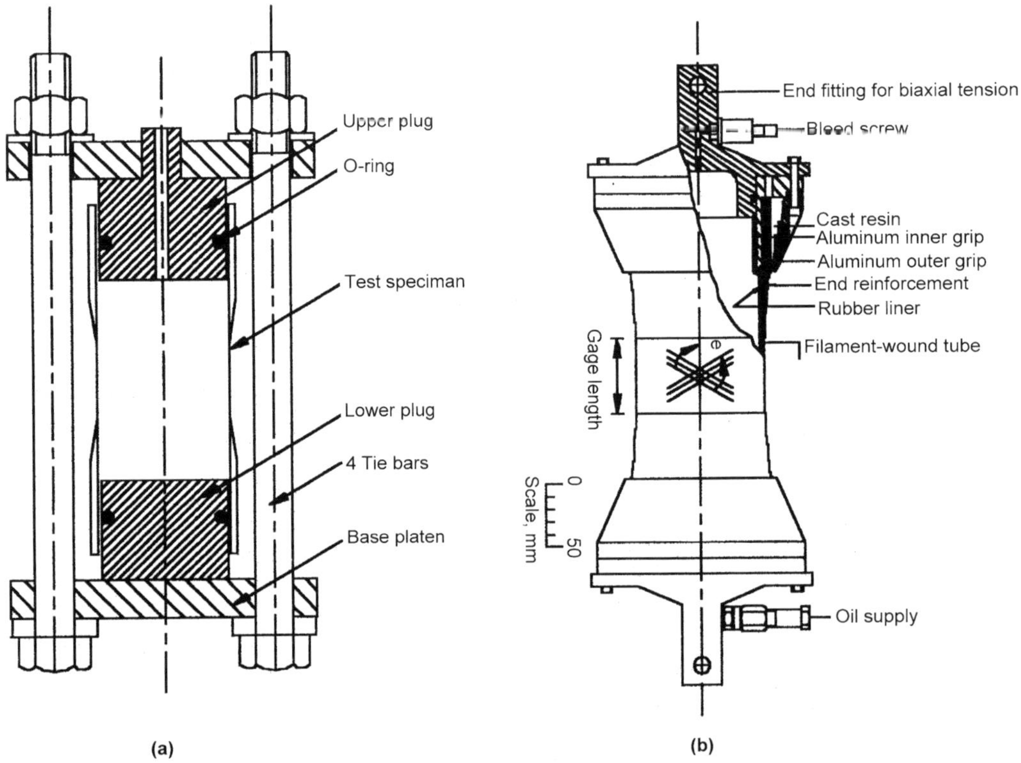


**Fig. 8.15** Test rigs used for biaxial tests. (a) Open-end burst. (b) For stress ratio = 2:1 and 1:1 and others. Source: Ref 8.67


8.9 "E-765 Approved Databases," Fibercote Industries, Waterbury, CT

8.10 *Annual Book of ASTM Standards*, American Society for Testing and Materials, Philadelphia, PA

8.11 "ISO Standards," International Organization for Standardization, http://www.iso.org/iso/iso_catalogue.htm

8.12 D. Cohen, Influence of Filament Winding Parameters on Composite Vessel Quality and Strength, *Compos. Part A: Appl. Sci. Manuf.*, Vol 28, 1997, p 1035–1047

8.13 M.F. Yokota, *SAMPE J.*, Vol 14 (No. 4), 1978, p 11

8.14 S.R. Ghiorse, Effect of Void Content on the Mechanical Properties of Carbon/Epoxy Laminates, *SAMPE Q.*, Jan 1993, p 54–59

8.15 J.J. Koppert and A.B. Beukers, Full Composite Isotensoid Pressure Vessels, or How Composites Can Compete with Steel, *SAMPE J.*, Vol 36 (No. 6), Nov/Dec 2000, p 8–15

8.16 Yu. M. Tarnopol'skii and V.L. Kulakov, in *Handbook of Composites*, 2nd ed., S.T. Peters, Ed., Chapman and Hall, London, 1998

8.17 C.T. Herakovich, *J. Compos. Mater.*, Vol 15 (No. 7), 1981, p 336

8.18 T.P. Philippidis, D.J. Lekou, and D.G. Aggelis, Failure of Filament Wound Structural Components Under Biaxial Loading, *Compos. Struct.*, Vol 45, 1999, p 41–50

8.19 D.W. Jensen and T.R. Pickenheim, "Compressive Behavior of Undulations in Filament Wound Composites," AIAA-93-1516-CP, 1993

8.20 A. Jones, V. Middleton, and M.J. Owen, Roller Assisted Variant of the Split Disk Test for Filament-Wound Composites, *Compos. Part A: Appl. Sci. Manuf.*, Vol 27, 1996, p 287–294

8.21 N.J. Pagano and J.M. Whitney, *J. Compos. Mater.*, Vol 4 (No. 7), 1970, p 360

8.22 N.J. Pagano, *J. Compos Mater.*, Vol 5 (No. 2), 1971, p 260

8.23 T. Dickson, S. Lee, and M. Munro, Development of a Vacuum Bagging Technique for Autoclave Curing of Filament Wound

Thin-Walled Tubes, *Mater. Manuf. Process.*, Vol 10 (No. 1), 1995, p 1–10
8.24 W.R. Swanson, G.R. Toombes, and S.W. Beckwith, In-Plane Shear Properties of Composites Using Torsion Tests of Thin-Wall Tubes, *29th National SAMPE Symposium*, April 3–5, 1984, p 567–577
8.25 B. Spencer and D. Hull, Effect of Winding Angle on the Failure of Filament Wound Pipe, *Composites*, Vol 9 (No. 1), Jan 1978, p 17–24
8.26 A. Onder, O. Sayman, T. Dogan, and N. Tarakcioglu, Burst Failure Load of Composite Pressure Vessels, *Compos. Struct.*, Vol 89, 2009, p 159–166
8.27 G. Nagy and S. Rosenwasser, The Evaluation and Testing of Graphite Materials for High Speed Rotors, *IEEE Trans. Magn.*, Vol 35 (No. 1), Jan 1999
8.28 C.W. Gabrys and C.E. Bakis, Design and Manufacturing of Filament Wound Elastomeric Composite Flywheels, *J. Reinf. Plast. Compos.*, Vol 16 (No. 6), 1997, p 408–502
8.29 J.C. Thesken, C.L. Bowman, S.M. Arnold, and R.C. Thompson, "Time-Temperature Dependent Response of Filament Wound Composites for Flywheel Rotors," NASA/TM 2004-212102
8.30 C.W. Gabrys and C.E. Bakis, Filament Winding of Thick Composite Rings with High Radial Accretion Rates, *Society of the Plastics Industry, Proceedings of the International Exposition*, May 31–June 4, 1998 (New York, NY), p 1904–1915
8.31 K. Takahashi, S. Kitade, and H. Morita, Development of High Speed Composite Flywheel Rotors for Energy Storage Systems, *Adv. Compos. Mater.*, Vol 11 (No. 1), 2002, p 41–50
8.32 M.M. Salama, D.B. Johnson, and J.R. Long, Composite Production Riser-Testing and Qualification, *SPE Prod. Facilities*, Aug 1998, p 170–177
8.33 J.R. Vinson and R.L. Sierakowski, *The Behavior of Structures of Composite Materials*, Martininus Nijhoff, Dordrecht, 1986
8.34 D. Cohen, Y.T. Toombes, H.K. Johnson, and M.F. Hansen, Pressurized Ring Test for Composite Pressure Vessel Hoop Strength and Stiffness Evaluation, *J. Compos. Technol. Res.*, Vol 17 (No. 4), Oct 1995, p 331–340
8.35 S. Arsene and J. Bai, A New Approach to Measuring Transverse Properties of Structural Tubing by a Ring Test, *J. Test. Eval.*, Vol 24 (No. 6), Nov 1996, p 386–391
8.36 D. Cohen, Influence of Filament Winding Parameters on Composite Vessel Quality and Strength, *Compos. Part A: Appl. Sci. Manuf.*, Vol 28, 1997, p 1035–1047
8.37 D. Kugler and T.J. Moon, A Technique for Compression Testing of Composite Rings, *Compos. Part A: Appl. Sci. Manuf.*, Vol 33, 2002, p 507–514
8.38 "Plastics Piping Systems—Glass-Reinforced Thermosetting Plastics (GRP) Pipes—Determination of Initial Specific Ring Stiffness," ISO 7685:1998
8.39 F. Pierron and P. Davies, Ring Compression Test for Cylindrical Composite Mechanical Characterization, *Fourth European Conference on Composites: Testing and Standardisation*, Aug 31–Sept 2, 1998 (Lisbon, Portugal), p 193–202
8.40 P. Samyn, L. Van Schepdael, W. Van Paepegem, J.S. Leedertz, E. Suister, P. DeBaets, P. Derieck, and J.S. Degrieck, Fracture Assessment of Carbon Fibre/Epoxy Reinforcing Rings Through a Combination of Full Scale Testing, Small Scale Testing, and Stress Modeling, *Appl. Compos. Mater.*, Vol 13, 2006, p 57–85
8.41 S.H. Yoon, C.G. Kim, and W.M. Cho, Measurement of Tensile Properties Using Filament Wound Ring Specimens, *J. Reinf. Plast. Compos.*, Vol 16 (No. 9), 1997, p 810–824
8.42 D. Cohen, Influence of Filament Winding Parameters on Composite Vessel Quality and Strength, *Compos. Part A: Appl. Sci. Manuf.*, Vol 28, 1997, p 1035–1047
8.43 S.K. Ha and J.Y. Jeong, Effects of Winding Angles on Through-Thickness Properties and Residual Strains of Thick Filament Wound Composite Rings, *Compos. Sci. Technol.*, Vol 65, 2005, p 27–37
8.44 C. Kaynak, E. Salim Erdiller, L. Parnas, and F. Senel, Use of Split-Disk Tests for Process Parameters of Filament Wound Epoxy Composite Tubes, *Polym. Test.*, Vol 24, 2005, p 648–655
8.45 P.C. Tse, S.R. Reid, and S.P. Ng, Spring Constants of Filament-Wound Composite Circular Rings, *Proc. Inst. Mech. Eng. C*, Vol 215 (No. 2), 2001, p 211–221

8.46 W.L. Ko and R.H. Jackson, "Open-Mode Delamination Stress Concentrations in Horseshoe and Elliptic Composite Curved Bars Subject to End Forces," NASA Technical Memorandum 4164, Ames Research Center, 1990

8.47 D.F. Adams, Through the Thickness Tensile Strength Testing Using a Curved Beam, *High-Perform. Compos.*, Nov 2008, p 11–13

8.48 A. Sharma and C. Bakis, C-Shape Specimen for Tensile Radial Strength of Thick Filament-Wound Rings, *J. Compos. Mater.*, Vol 40 (No. 2), 2006, p 97–115

8.49 R.F. Foral, D.D. Baldwin, and J.M. McGee, in *Proceedings of the Spring Conference on Experimental Mechanics*, Society of Experimental Mechanics, 1986, p 399–405

8.50 D.L. Smith and M.B. Dow, "Properties of Three Graphite/Toughened Resin Composites," NASA Technical Paper 3102, 1991

8.51 T. Messager, M. Pyrz, B. Gineste, and P. Chauchot, Optimal Laminations of Thin Composite Cylindrical Vessels, *Compos. Struct.*, Vol 58 (No. 4), Dec 2002, p 529–537

8.52 D.E. Walrath and D.F. Adams, The Iosipescu Shear Test as Applied to Composite Materials, *Exp. Mech.*, March 1983, p 105–110

8.53 K.T. Kedward, On the Short Beam Test Method, *Fibre Sci. Technol.*, Vol 5, 1972, p 85–95

8.54 D.F. Adams, A Comparison of Shear Test Methods, *High-Perform. Compos.*, Sept 2005, p 9–10

8.55 D.F. Adams and J.M. Busse, Suggested Modifications of the Short Beam Shear Test Method, *International SAMPE Symposium*, May 16–20, 2004

8.56 N.L. Newhouse and W.D. Humphrey, Development of the Standard Test Evaluation Bottle (STEB), *17th National SAMPE Technical Conference*, Oct 22–24, 1985, p 554–562

8.57 W.D. Humphrey and N.L. Newhouse, The Standard Test and Evaluation Bottle (STEB), Five Years Later, *31st International SAMPE Symposium*, April 7–10, 1986, p 1383–1394

8.58 Y.K. Huang, P.H. Frings, and E. Hennes, Exploding Pressure Vessel Test on Zylon/Epoxy Composite, *Compos. B: Eng.*, Vol 33, 2002, p 117–123

8.59 T.K. Hwang, C.G. Hong, and C.G. Kim, Size Effect on the Fiber Strength of Composite Pressure Vessels, *Compos. Struct.*, Vol 59, 2003, p 489–498

8.60 D. Cohen, S. Mantell, and L. Zhao, The Effect of Fiber Volume Fraction on Filament Wound Composite Pressure Vessel Strength, *Compos. B: Eng.*, 2001, p 413–429

8.61 J.S. Park, C.S. Hong, C.G. Kim, and C.U. Kim, Analysis of Filament Wound Composite Structures Considering the Change of Winding Angles Through the Thickness Direction, *Compos. Struct.*, Vol 55, 2002, p 63–71

8.62 T.K. Hwang, Y.D. Jung, W.M. Doh, and B.J. Cho, The Performance Improvement of Filament Wound Composite Pressure Vessels, *45th International SAMPE Symposium*, 2000, p 1427–1438

8.63 J.-S. Park, C.-U. Kim, H.-K. Kang, C.-S. Hong, and C.-G. Kim, Structural Analysis and Strain Monitoring of the Filament Wound Motor Case, *J. Compos. Mater.*, Vol 36 (No. 20), 2002, p 2373–2388

8.64 Y. Doh and C. Hong, Progressive Failure Analysis for Filament Wound Pressure Vessel, *J. Reinf. Plast. Compos.*, Vol 14, Dec 1995, p 1278–1305

8.65 P.D. Soden, M.J. Hinton, and A.S. Kaddour, Recommendations for Designers and Researchers Resulting from the World-Wide Failure Exercise, *Compos. Sci. Technol.*, Vol 64, 2004, p 589–604

8.66 P. Soden, Predicting the Strength of Composite Laminates, *Reinf. Plast.*, Feb 2005, p 42–48

8.67 P.D. Soden, M.J. Hinton, and A.S. Kaddour, Biaxial Test Results for Strength and Deformation of a Range of E-Glass and Carbon Fibre Reinforced Composite Laminates: Failure Exercise Benchmark Data, *Compos. Sci. Technol.*, Vol 62, 2002, p 1489–1514

8.68 M.J. Hinton, A. Kaddour, and P.D. Soden, *Failure Criteria in Fibre-Reinforced Polymer Composites*, Elsevier, London, U.K., 2004

Composite Filament Winding
S.T. Peters, editor

Copyright © 2011, ASM International®
All rights reserved.
www.asminternational.org

# CHAPTER 9

# Pressure Vessel Design, Fabrication, Analysis, and Testing

Hugh Reynolds

## Introduction

Composite filament-wound pressure vessels were begun for solid rocket motor cases (RMCs), which were primarily responsible for accelerating filament winding from a laboratory curiosity to the major industry it is today. In solid rocket motors, the case or combustion chamber is a major part of the rocket motor structure and inert weight. The necessity of developing the lightest-weight structures was the driving factor for the development of filament-wound composite pressure vessels. In this chapter, a brief history is presented, followed by a comparison of the advantages of composites over metals for RMCs. A discussion of a typical design, analysis, and manufacturing operation follows. Finally, a solved problem illustrating the design techniques used for preliminary design is presented along with a discussion on demonstrated performance.

## History

If necessity is the mother of invention, then solid RMCs (combustion chambers) are the mother of the filament-winding industry. The process was first recorded during World War II in support of the Manhattan nuclear project, where lightweight plastic hoops were required (Ref 9.1). These were made on a lathe by George Lubin and Walt Greenberg of Bassons Industries. These hoop windings eventually became the basis for the famous Naval Ordnance Laboratory rings, which have since become the industry standard for short-beam shear and tensile ring testing.

Dick Young was one of the early leaders who filed key patents in this fledgling technology and then started his own company. His development work included filament-wound pipe nozzles and prototype RMCs. He licensed this technology to several other manufacturers and eventually sold out to Hercules Powder Company.

Rocket motor case technology came of age in the early 1960s. Participants included Thiokol, Hercules, Brunswick, Aerojet, Hitco, United Technology, and several others. Development of filament-wound cases for solid rocket motors was well supported by the Air Force. The Minuteman and Polaris rocket motors were among the first to be built in production quantities using filament winding. Design tools and manufacturing methods jointly evolved to bring filament winding into wide acceptance.

Today, solid rocket motors continue to use filament-wound pressure vessels for combustion chambers in Peacekeeper (stages 1, 2, and 3), Trident I and II (stages 1, 2, and 3), commercial launch vehicles such as Delta II and IV GEM-40, 46, and 60 strap on, Atlas V strap on, Taurus Castor 120, Pegasus Orion, Ground-Based Midcourse Defense (stages 1 and 2), Kinetic Energy Interceptor (stages 1 and 2), and several tactical motors such as Pac-3 and Thaad. The field has also expanded to include commercial pressure vessels for compressed natural gas (CNG), propane, compressed hydrogen fuel tanks, fireman air tanks, liquid fuel tanks, and even cryogenic fuel and oxygen tanks. Fuel tank plumbing also uses filament-wound composite pressurized cylindrical tubing (pressure vessel without the domes). Typically, commercial composite overwrapped pressure vessels (COPVs) are wound over a metallic or plastic liner that often serves as the winding mandrel.

## Why Composites

The Cold War was the catalyst, and long-range strategic RMCs (combustion chambers for the solid rocket motors) were the immediate objective. Every pound that could be eliminated from the RMC could be used to increase the range or payload of the missile. Some of the benefits of filament winding include:

- The primary reason for using composites is their higher strength-to-weight ratios over those of isotropic materials. The performance of RMCs and pressure vessels can be ranked by using the performance factor (N). It is sometimes called pressure vessel efficiency:

$$N = \frac{PV}{w} \qquad \text{(Eq 9.1)}$$

where $P$ is the pressure contained (kg/cm$^2$), $V$ is the internal volume (m$^3$), and $W$ is the weight (kg).

The weight component, $W$, can be used to include the weight of the entire RMC assembly, that is, nozzles, skirts, igniter, insulation, and the case. However, considering the weight of the composite pressure vessel case alone is more generally used when making material selections. Here, the high strength-to-weight ratio of composites (Table 9.1) clearly shows their advantages (Ref 9.2).

- Figure 9.1 plots some typical performance factors ($N \times 10^{-6}$). The performance factor is a figure of merit used to compare the expected or actual performance achieved in solid RMCs. The data illustrate how fiber properties (strength-to-density ratio) dominate this parameter. The data also show that the geometry of the case (length to diameter) has little effect on the performance factor.
- The versatility of filament winding saves weight by tailoring the winding patterns to carry the loads. This is particularly advantageous in closed-end pressure vessels where the longitudinal loads are only half that of the hoop loads. Structures made of isotropic materials are less efficient because the longitudinal stresses use only half that of the structure capabilities. Thinner wall sections are therefore inherent with composite cylinders, as are lower densities for the structure.
- The entire tooling package required for filament winding is considerably cheaper than for metal cases, plus the lead times required to procure these tools are much shorter.
- Alterations to the mandrel and other fabrication tools are far easier to implement; thus, design changes are easily implemented.
- Composites are basically unaffected by stress-corrosion cracking; however, metals are severely limited. Stress corrosion can result in spontaneous failures without warning or any evidence of impending failures. Safety in metal RMCs is generally addressed by assigning higher safety factors than normally used in composites.
- The low cost of the raw materials used in filament winding is another real advantage to composites. High-nickel steels, maraging steels, forgings, and welding operations involve considerable expense with long lead times. On the other hand, fiberglass roving and the resin components are inexpensive. Some prepreg roving systems that are

**Table 9.1 Comparison of motor case materials**

| | D6AC steel | Maraging steel | Titanium (6%Al-4%V) | Glass-filament composite (S-2)(a) | Organic-filament composite (Kevlar)(b) | Intermediate-modulus polyacrylontrile carbon fiber |
|---|---|---|---|---|---|---|
| Tensile strength, ksi (MPa) | 230 (1586) | 250 (1724) | 160 (1103) | 170 (1172) | 120 (827.4) | 225 (1551) |
| Density, lb/ft$^3$ (g/cm$^3$) | 489 (7.883) | 499 (8.00) | 124 (4.429) | 124 (1.993) | 85 (1.356) | 97 (1.550) |
| Strength/density | 202.5 | 215.5 | 249.0 | 588.1 | 610.2 | 1000.6 |
| Modulus, $10^6$ psi, (GPa) | 29 (199.5) | 27 (189.1) | 16 (110.3) | 5 (31.72) | 11 (75.84) | 25 (172.4) |

(a) Typical Glass-filament-wound motor case with approximately 80% Owens Corning (S-2) glass (600 ksi, or 4137 MPa, tensile strength) and 20% epoxy resin matrix. (b) Typical organic-filament-wound motor case with approximately 70% DuPont Kevlar 49 (500 ksi, or 3447 MPa, tensile strength) filament and 30% epoxy resin matrix

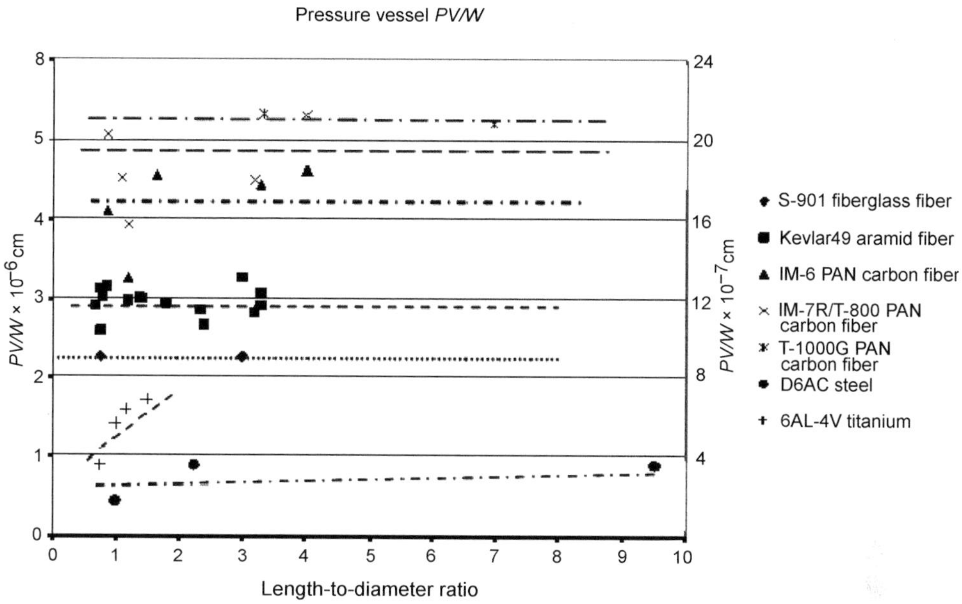


**Fig. 9.1** Typical performance factors. *P*, pressure; *V*, volume; *W*, weight; PAN, polyacrylonitrile

required by specifications can be relatively expensive; however, wet winding with resins formulated and mixed in-house by the manufacturer presents a cost-effective alternative to steel motor cases.

## Design Considerations

Over the years, the design of RMCs has evolved to a more exact science, using the latest computer analysis and computer-controlled manufacturing methods. A program originally developed for mainframe computers and now available for personal computers has been developed under Air Force funding, called the Composite Design and Analysis Code (Ref 9.3), where the entire composite pressure vessel, composite skirts, and associated hardware (polar boss, O-rings, closure, and fasteners) and attachments (attach flanges and fasteners) can be designed by the operator at the computer station (including solid rocket motor combustion cases and commercial pressure vessels, including COPVs of every configuration). The current version of this program is available from the Air Force Rocket Propulsion Laboratory and runs on a typical personal computer. Users input volume, envelope, weight, and pressures along with material properties, and the program outputs iterative solutions for review and adjustment. Then, a more complete analysis may be performed using any of several finite-element analysis tools that are commercially available. Comparisons between the predicted and actual design details and performance have been quite good. Access to these analytical tools is not always available to the everyday composite designer. However, quick sizing estimates and rough design calculations can be valuable tools. The following paragraphs introduce the basic design approach and show some sizing techniques along with example calculations.

### *Nomenclature*

The typical nomenclature for a pressure vessel is shown here and pictorially in Fig. 9.2. It consists of geometric parameters such as lengths, diameters, opening sizes, and interfaces with other structures; design features such as skirt attachments and skirt-to-pressure-vessel structural Y-joints, flanges, and polar fittings; the definitions of hoop and helical wind angles; and what constitutes a ply and a layer for both hoop and helical windings. This nomenclature applies to all pressure vessels, including those used in the fuel tank industry for natural gas, propane, air, hydrogen, and so on.

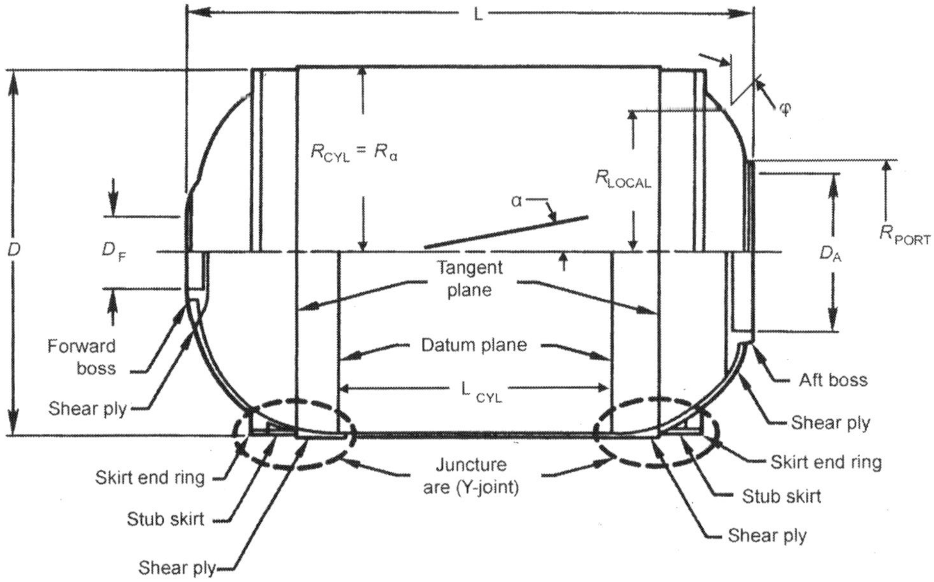


Composite case nomenclature

- Fwd and aft port openings ($D_F$, $D_A$)
- End fitting or polar boss (opening attach feature for nozzle/igniters)
- Diameter (D): Twice the radius (either local or cylinder)
- Overall length (L = OAL)
- Cylinder length ($L_{CYL}$)
- Tangent line ($T_L$): Dome cylinder junction
- Datum plane (drawing datum for defined features)
- Skirt (means of attaching the pressure vessel to other structure)
- Y-joint (skirt-to-pressure-vessel joint shear ply)
- Skirt flange (attach feature)
- End fitting shear ply (allows differential movement of boss and windings)
- Skirt-to-pressure-vessel design concept: Integral or secondary attach
- Hoop angle (fiber angle near 90° to axis of motor)
- α is the local helical or polar angle (fiber angle to the axis of the pressure vessel)
- Discrete number of layers
  a. One helical = 2 plies (±α)
  b. One hoop layer = 2 plies (one complete pass down and back along cylinder)

**Fig. 9.2** Composite pressure vessel nomenclature

### *Requirements*

The design is initiated with typical customer- or specification-defined design requirements. These types of requirements typically define geometry limits, limit or ultimate structural loads, stiffness (axial, bending, shear, torsion), first bending mode natural frequency, safety factors, requirements for materials used, environmental requirements such as temperature, humidity, and life, as well as overall motor reliability, which is then broken down by the contractor into the requirements for each motor component, as shown in Table 9.2.

The contractor adds additional requirements that are necessary to define the product—the pressure vessel or insulated motor case. Typical contractor requirements define operational conditions, such as internal maximum expected operating pressure, maximum motor thrust (if applicable), internal volume, materials selection, design parameters such as minimum burst pressure, design fiber strengths, allowable fiber strain, and so on. Because of structural considerations, the skirt-to-case-joint design may also be specified, usually based on trade studies conducted by the contractor, which lean heavily on past experience with similar designs.

## *Materials Selection*

Materials selection is a key parameter for designing a pressure vessel. Typically, this starts with fiber selection. Fiber selection is usually based on performance needs such as burst strength, weight, or stiffness requirements. In the 1950s and 1960s, the only fibers available were the E- and S-glass inorganic fibers. In the 1970s, the organic aramid fiber, Kevlar 49 (DuPont), became available, with higher specific stiffness compared to glass fibers. In the late 1980s and most of the 1990s, the fibers of choice were intermediate-modulus, high-strength polyacrylonitrile-based carbon fibers such as T800 and IM7, both 40 to 43 × $10^6$ psi (276 to 294 GPa) modulus and 800 ksi (5516 MPa) tensile strength fibers available at $60/lb. Currently, the fibers of choice are M30SC or T1000GB. Where high performance at low cost is desired, M30SC is the top choice, with 800 ksi (5516 MPa) tensile strength and 43 × $10^6$ psi (294 GPa) modulus available at $29/lb. Where premium strength is required, the T1000GB fiber is one of the strongest available at 920 × $10^6$ psi (6370 GPa) with a modulus of 43 × $10^6$ psi (294 GPa) at a cost of $85/lb. A relatively new fiber of interest is the Toho G40-800 12,000 and 24,000 filament tow, at 42 × $10^6$ psi (290 GPa) modulus with tensile strength of 850 ksi (5861 MPa) and prices in the $30 to $50/lb range. Another fiber of interest is organic poly(p-phenylenebenzobisoxazole) (PBO) fiber, which offers the strength and modulus of intermediate-modulus, high-strength carbon fiber with the toughness of aramid fibers. Use of PBO fibers in RMC applications has been limited to date due to some of the poor shear and compressive strength limitations common to the aramid fibers, such as Kevlar 49. An emerging organic fiber is M5, with a target tensile strength of more than 1.3 × $10^6$ psi (8963 MPa), a modulus of 50 to 65 × $10^6$ psi (345 to 448 GPa), and attractive compressive strength capability to 435 ksi (3000 MPa). This material is not yet available in sufficient quantities to verify the manufacturer's claims. Table 9.3 shows the types and characteristics of fibers available today.

The next most important material is the resin selection. A variety of resin systems have been used for making composite motor cases. Widely

**Table 9.2 Motor component design requirements**

| Parameters | Typical values |
|---|---|
| Geometry: Interface control drw (ICD) | |
| Length | 80 in. (2032 mm) |
| Diameter | 92 in. (2337 mm) |
| Ultimate interface structural strength (line load) | |
| Compression | –3000 lb/in. (–525 kN/m) |
| Tension | +2700 lb/in. (+470 kN/m) |
| Shear and torsion | ±2000 lb/in./ (±350 kN/m) |
| Cylindrical stiffness | |
| Bending stiffness (EI) | 10 × $10^{11}$ lbf/in. (2.87 × $10^9$ N/M$^2$) |
| Factors of safety | |
| Proof | 1.05 |
| Flight | 1.25 |
| Captive flight | 1.4 |
| Materials allowable (Mil-HDBK-5, 17, 23, etc.) | |
| Metallic | A-basis |
| Composite | B-basis |
| Environments | |
| Temperature | 24.8 to 102 °F (–4 to 39 °C) |
| Humidity, %RH | Up to 100 |
| Life | >10 years |
| Reliability | 0.999 |

**Table 9.3 Fiber characteristics for available inorganic and organic fibers**

| Fiber type | Tensile modulus, $10^6$ psi (GPa) | Tensile strength, ksi, (MPa) | Strain to failure, % | Fiber density, lb/ft$^3$ (g/cm$^3$) | Cross-sectional area, in.$^2$ (mm$^2$) | Supplier/product |
|---|---|---|---|---|---|---|
| E-glass | 10.5 (72.5) | 500 (3447) | 4.8 | 163 (2.6) | 12 × $10^{-4}$ (0.134) | Owens Corning |
| S-glass | 12.6 (86.9) | 650 (4482) | 5.3 | 159 (2.55) | 12 × $10^{-4}$ (0.134) | Owens Corning |
| Aramid | 175 (121) | 550 (3792) | 3 | 89.91 (1.44) | 5 × $10^{-4}$ (0.352) | Devlar 49 DuPont |
| Oriented polyethylene | 24.9 (172) | 435 (3000) | 1.9 | 60 (0.970) | 4 × $10^{-4}$ (0.298) | Sectra 1000 Honeywell |
| PBO | 39.1 (270) | 836 (5764) | 1.9 | 97 (1.56) | 3 × $10^{-4}$ (0.2097) | Toyobo |
| High-strain PAN carbon | 35.2 (243) | 730 (5030) | 2.1 | 111.7 (1.79) | 1.4 × $10^{-4}$ (0.914) | G30-700 Toho |
| | 33.3 (230) | 711 (4900) | 2.1 | 112.3 (1.8) | 1.4 × $10^{-4}$ (0.917) | T-700S Toray |
| Intermediate-modulus PAN carbon | 40 (276) | 808 (5570) | 1.90 | 111.1 (1.78) | 4 × $10^{-4}$ (0.25) | IM-7R Hexcel |
| | 42.6 (294) | 796 (5490) | 1.90 | 108 (1.73) | 6.8 × $10^{-4}$ (0.4393) | M30SC Toray |
| | 41.9 (289) | 825 (5690) | 1.97 | 111.7 (1.79) | 7 × $10^{-4}$ (0.464) | G40-800 Toho |
| Intermediate-modulus high-strain PAN carbon | 44 (304) | 891 (6140) | 1.9 | 112.3 (1.8) | 2.8 × $10^{-4}$ (0.186) | IM-9 Hexcel |
| | 42.6 (294) | 924 (6371) | 2.2 | 112.3 (1.8) | 4 × $10^{-4}$ (0.269) | T1000GB Toray |
| High-modulus PAN carbon | 63.2 (436) | 611 (4210) | 1 | 114.9 (1.64) | 3.8 × $10^{-4}$ (0.242) | M-46J Toray |

PBO, poly(p-phenylenebenzobisoxazole); PAN, polyacrylonitrile

used resins include heat-cured epoxies with temperature capabilities of approximately 249 to 320 °F (120 to 160 °C). Resins in this class can be wet wound (in-process impregnated, usually mixed by the user just prior to use), stage rewound (wet prepreg), or prepreg wound (preimpregnated tow respooled and delivered frozen to the customer by specialty prepreggers). The listed resins are just a small fraction of the resins used but represent the various temperature and processing classes in current production today. Typically, the higher the temperature capability for a resin, the more difficult it is to process and the more costly to manufacture. For instance, fabricating a multilayered motor case with polyimide resins costs five times as much in labor compared to a conventional wet-wound epoxy because it must be imidized between layers to avoid condensation-induced fiber wrinkling. Typical resin types available today are shown in Table 9.4.

The most common resins used for filament-wound pressure vessels are the wet winding resins in the 249 to 320 °F (120 to 160 °C) capability class. Some representative types are shown in Table 9.5. Some of these have fallen victim to environmental restrictions on the use of suspect carcinogens. Methylene dianiline (MDA) and eutectic blends using MDA have become restricted or eliminated from use. Alternatives are typically closely guarded proprietary formulations, so the examples listed here are out of date but representative of this class of resins and their characteristics.

### *Design Approach*

Hoop windings cover the cylindrical section of the pressure vessel and are called circumferential windings. Longitudinal or helical wind-

**Table 9.4 Common resin types**

**Resins utilized (representative examples)**

**Winding resin systems**

- Moderate temperature (capability 248–320 °F, or 120–160 °C), ambient wind
  LRF-092: Bis A/anhydrided/amine catalyzed
  LRF-205/UTR72-01: Bis A/cycloaliphatic/aromatic amine
  HBRF-55: Bis A/butanediol/aromatic amine
  HBRF-241: Bis A/butanediol/flexible/aromatic amine
  UTR99-01: Bis A/butanediol/aromatic amine
  UTR00-01A-3: Cycloaliphatic/aromatic amine/amine catalyzed
- High temperature (capability 347–410 °F, or 175–210 °C), warm wind (up to 131 °F, or 55 °C)
  HBRF-295: Epoxy novalac/aromatic amine
  UTR89-14A: Epoxy novalac/cycloaliphatic/aromatic amine
  UTR99-16D: Multifunctional/aromatic amine/amine catalyzed
- Very high temperature (capability 419–788 °F, or 215–420 °C), hot wind (up to 212 °F, or 100 °C)
  PRM resins: Cytec Cycom 944, 966D (PMR-15), 3002, 5004
  BMI resin: Cytec Cycom 5250, 5260, 5392
  Cyanate ester: BA-200 (Lonza)
  Phenolic triazine resins: PT-30 (Lonza)

**Prepreg systems**

- Cytec ERL 1908: Bis A/cycloaliphatic/thermoplastic/aromatic amine
- TCR UF3323: Long outlife at room temperature (12 mo.) dual cure
- Cytec Cycom 977-2-epoxy, Cycom 250-4-BMI, Cycom 5245C-eposy/cyanate ester, Cycom 944, 966D, 3002, 5004-PRM
- EF-2: Bis A/Multifunctional/anhydride/amine catalyzed

**Table 9.5 Wet winding resins and characteristics for representative epoxies**

| Resin | A | B | C | D | E | Supplier |
|---|---|---|---|---|---|---|
| Epon 828 | 65 pbw | ... | 100 pbw | 65 pbw | ... | Hexion |
| Epon 826 | ... | 100 pbw | ... | ... | ... | Hexion |
| Epolite 2447 | ... | ... | ... | ... | 100 pbw | Hexcel |
| ERL 4206 | 35 pbw | ... | ... | 35 pbw | 5 pbw | Union Carbide |
| Heloxy 67 | ... | 25 pbw | ... | ... | ... | Hexion |
| MDA | 39.5 pbw | ... | ... | ... | ... | Japan |
| Tonox 6040 | ... | 30 pbw | ... | ... | ... | Japan |
| Epolite 2330 | ... | ... | ... | 21.6 pbw | 23 pbw | Hexcel |
| NMA | ... | ... | 90 pbw | ... | ... | ... |
| BDMA | ... | ... | 1 pbw | ... | ... | ... |
| Viscosity, cP | 560 | 1200 | 1500 | 560 | 520 @ 120 °F (49 °C) | ... |
| Pot life, h | 18 | 7 | 72 | 17 | 4 | ... |
| Tensile strength, ksi, (MPa) | 141.1 (97.2) | 12 (82.7) | 9 (62) | 12 (82.7) | 5 (34.5) | ... |
| Modulus, ksi (GPa) | 520 (3.58) | 421 (2.90) | 500 (3.45) | 500 (3.45) | 544 (3.75) | ... |
| Density, lb/ft$^3$ (g/cm$^3$) | 75 (1.202) | 75.7 (1.213) | 74.9 (1.2) | 75 (1.202) | 77 (1.235) | |
| $T_g$, °F (°C) | 329 (165) | 289.4 (143) | 293 (145) | 820 (160) | 406.4 (208) | |

pbw, parts by weight; MDA, methylene dianiline; NMA, nadic methyl anhydride; BDMA, benzyl dimethylamine; $T_g$, glass transition temperature

ings cover the cylinder and dome sections of the pressure vessel. Generally, there are two types of helical or longitudinal fiber windings: polar winding and helical winding. Both are controlled by the available geometry. Polar winding is geometrically limited to certain relationships defined by the port openings and the length of the part. Helical winding is more versatile in that greater differences in geometry can be accommodated by helical winding than with polar winding. However, helical winding requires more sophisticated winding machines. Wind ability is determined during wind trials by the deviation of the actual fiber path from the ideal zero friction or geodesic path. The path deviation can be caused by fiber slippage, improper dome contours, or incorrectly programmed winding machine error. For a specific fiber/resin combination, there exists a limit to the path deviation that can be tolerated, and this is called slip angle or friction limit. Generally, these limits are determined empirically.

Dome shapes are derived by analytical models that consider the fiber path with or without resin and whether there are allowed deviations from the geodesic path. The most common dome types are the Nourse dome (a netting-derived dome where the resin modulus is neglected and the wind angle is preselected with or without shaping factor; local angles must be estimated using friction models or semiempirical correlations to measured techniques); the balanced geodesic dome (fixed derivable wind angle without fiber slippage without shaping factor), which is rarely used on rocket motors but is common on commercial pressure vessels such as fireman air tanks, CNG, and fuel tanks (also known as COPVs); balanced modified geodesic (selected wind angle with or without shaping factor considering port opening and allows deviation from geodesic; local angles by semiempirical or friction path defined; resin modulus can be included, similar to Nourse); and the zero hoop stress (a special case of Nourse dome with zero wind angle used to generate; local angles must be estimated using friction models or semiempirical correlations to measured techniques). Another dome contour that is widely used for relatively short pressure vessels is the polar dome (cylinder angle defined strictly by length, diameter, and port openings; local angles defined by geometry-provided deviations from geodesic can be tolerated). The most general nonpolar dome is the unique balanced ovaloid shapes, depending on wind method, helical angle, length, diameter, polar opening, and fiber path friction considerations. This latter dome type includes the Nourse, geodesic, and modified geodesic but adds a friction-controlled fiber path over the entire wound surface (i.e., the wind angle continuously changes).

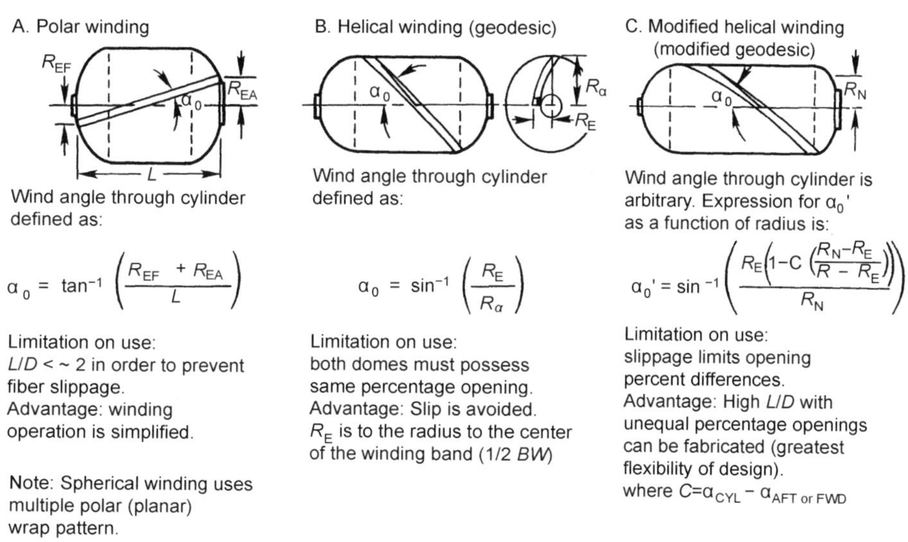


**Fig. 9.3** Basic design approach for typical domes

All dome shapes can be derived by assuming either a starting helical angle for a given port size or deriving a starting angle based on the motor case geometry. Domes can also be shaped by forces on the apex with a shaping factor, $\tau$, which is a dimensionless force. Local wind angles are estimated by semiempirical correlations to measured techniques or angles and can be calculated using dome shape and fiber path friction models. The basics of wind angles and dome shapes are shown in Fig. 9.3. This figure covers the general approaches for the three basic types: polar, geodesic, and modified geodesic (also known as Nourse, balanced modified geodesic, or balanced ovaloid).

Cylinders are just that, cylindrical shapes, although conical sections of pressure vessels have been wound successfully with cone angles up to 4° and higher with the appropriate fiber/resin combination that allows large geodesic path deviations.

Structural attachments or skirts are typically hand or machine lay-up (fiber placement) of axial (longitudinal) and ±45° to axis reinforcements (unidirectional or woven) coupled with hoop or circumferential windings. Skirts may be cylindrical (most common) or conical (requiring more complex manufacturing and tooling techniques).

The guidelines for selecting a starting wind angle for the cylinder are shown in Table 9.6. The slippage allowed is the basis for selection. This chart gives some guidelines on selecting a starting $\alpha$. As the procedure continues, the winding machine requirements must be considered so that the angle and path selected is achievable with the available equipment.

**Netting Analysis.** The netting analysis (Ref 9.4) assumes that the fibers provide all the longitudinal stiffness and strength. This assumption is not only conservative but also an excellent basis for quick sizing of composite thickness of a typical RMC or pressure vessel where the liner does not have a significant load share. These equations have been derived in Ref 9.4 to 9.6. A quick review is presented here.

In the cylindrical region of an RMC or pressure vessel, the helical fibers carry the entire longitudinal load. A free-body diagram for a pressure vessel is shown in Fig. 9.4. Summing the running loads in the longitudinal direction, $N_\varphi$, the following relationships are obtained.

For the purpose of this discussion, a typical ply thickness, $T$, for most applications is:

$$T_{REF} \cdot CSA/CSA_{REF} \qquad \text{(Eq 9.2)}$$

where CSA is the cross-sectional area, and $T_{REF}$ = 0.006 in. (0.142 mm) for a $CSA_{REF}$ of 0.0004 in.$^2$ (0.246 mm$^2$).

A layer is then 2 times the ply thickness, or 0.011 in. (0.284 mm). This approach is a rule-of-thumb technique that has its basis in demonstrated experience. For example, for large-filament-count tows, the ply thickness is approximately (6.8122/3.815) × 0.0056 in., or 0.010 in. ([0.439/0.246] × 0.142 mm, or 0.254 mm). (This is

**Table 9.6 Design approach for wind angle selection criteria**

- Calculate $\alpha$ for cylinder
  Geodesic $\alpha_{AFT} = \sin^{-1}((D_A + BW)/(D_{CYL}))$
  Geodesic $\alpha_{FWD} = \sin^{-1}((D_F + BW)/(D_{CYL}))$
  Modified geodesic $\alpha_{CYL} \approx (\alpha_{FWD} + \alpha_{AFT})/2$ or $(\alpha_{FWD} + 2 \cdot \alpha_{AFT})/3$
  Planar (polar) $\alpha_{CYL} = \tan^{-1}((D_F + D_A + 2 \cdot BW)/(2 \cdot L))$
- Look at the devaiations from geodesic
  $\text{Abs}(\alpha_{CYL} - \alpha_{AFT}$ (or $\alpha_{FWD})) < 8$–10° 5–6° for graphite or carbon fiber)
- Are dome openings compatible?
  $(\alpha_{AFT} - \alpha_{FWD}) < 15°$ (10–12° for graphite or carbon fiber)
- Special considerations
  Select $\alpha_{CYL}$ for helical for long cylinder
  a. $\alpha_{CYL}$ may be varied in cylinder if sufficient cylinder length is available as follows:
  Where $1/(\sin(\alpha_{AFT})) - 1/(\sin(\alpha_{FWD})) = k \cdot L_{CYL}/R_{CYL}$, where k is the cylinder fiber path friction coefficient ~ 0.10 for wet impregnated carbon fibers

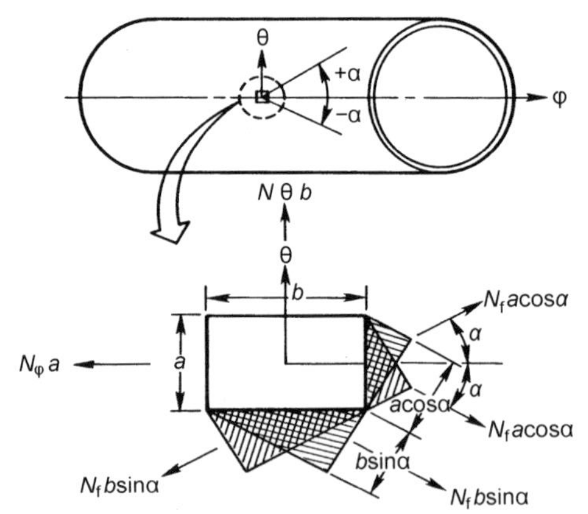


**Fig. 9.4** Netting analysis free-body diagram

not a hard-and-fast rule but useful for most applications.) The designer normally has a maximum ply thickness in mind when starting the design, based on prior experience with the materials. For the following discussion, assume a ply thickness of approximately 0.010 in. (0.254 mm).

For a pressure vessel with closed ends, the longitudinal running load, $N_\varphi$, can be defined in terms of the internal pressure, $P$, and the inner wall radius, $R_{CYL}$. The circumferential running line load, $N_\theta$, can also be defined in terms of the internal pressure and the inner wall radius. The more accurate estimates that agree with measured pressure vessel response use a correction for the inner wall radius by including the hoop strain. This requires a more sophisticated computation by rearranging the terms in the equations but is well within most calculator and personal computer software capabilities. The fiber characteristics needed to compute the number of winding layers required to achieve the necessary thickness include the fiber modulus and cross-sectional area. Both of these parameters were previously tabulated for common fibers in Table 9.3 in the section "Materials Selection:"

$$N_\varphi\ (\text{axial}) = P \cdot R_C/2 - \text{Thrust}/(2\pi(R_C + \text{THICK}/2)) + \text{Moment}/(\pi(R_C + \text{THICK}/2)^2) \quad \text{(Eq 9.3)}$$

where $P$ is internal pressure; $R_C = R_{CYL} \cdot (1 + \varepsilon_{Hoop})$ is the cylinder inside radius adjusted for hoop strain; "Thrust" is the motor thrust (at $P$); "Moment" is the added maximum external cylinder bending moment at the carrying internal pressure, $P$; $\varepsilon_{Hoop}$ is the cylinder membrane hoop strain ($\sigma_{HOOP}/E_{FIBER}$); and $E_{FIBER}$ is the fiber tensile modulus.

The helical composite thickness is derived solely from the axial line load, $N_\varphi$, the $\cos^2$ function of the cylinder wind angle, and the allowed helical fiber strength and fiber volume:

$$\text{THICK}_{HELICAL} = N_\varphi / (VF_{HELICAL} \cdot \cos^2(\alpha_{CYL}) \cdot \sigma_\alpha) \quad \text{(Eq 9.4)}$$

where $\alpha_{CYL}$ is the cylinder wind angle, $\sigma_\alpha$ is the allowable helical fiber strength, and $VF$ is the fiber volume (fraction of fiber volume to composite volume).

For the helical composite:

$$VF_{HELICAL} = 1/(1 + (W_R/(1 - W_R)) \cdot \rho_F/\rho_R) \quad \text{(Eq 9.5)}$$

where $W_R$ is the resin weight fraction, and $\rho_F$ or $\rho_R$ is the fiber or resin density. The number of helical layers equals the integer of $\text{THICK}_{HELICAL}$/0.020 in. (0.508 mm), and the number of plies is 2 times the number of layers:

$$N_\theta\ (\text{hoop}) = PR_C \quad \text{(Eq 9.6)}$$

The hoop fiber composite thickness in the cylindrical region is derived partly by the hoop load, $N_\theta$, partially by the axial load, $N_\varphi$, and the allowed hoop fiber strength and fiber volume. Considering the free-body diagram, it is seen that the contribution of the helical fibers in carrying the hoop load is a $\tan^2$ function of the cylinder wind angle times the axial line load, $N_\varphi$:

$$\text{THICK}_{HOOP} = (N_{HOOP} - N_\varphi \cdot \tan^2(\alpha_{CYL})) / VF_{HOOP}/\sigma_{HOOP} \quad \text{(Eq 9.7)}$$

where $\sigma_{HOOP}$ is the allowable hoop fiber strength, and $VF_{HOOP}$ is the hoop composite fiber volume and calculated as shown previously. The number of hoop layers is the integer of $\text{THICK}_{HOOP}$/0.020 in. (0.508 mm), and the number of plies is 2 times the number of layers.

For cylinder thickness:

$$\text{THICK}_{COMPOSITE} = \text{THICK}_{HELICAL} + \text{THICK}_{HOOP} \quad \text{(Eq 9.8)}$$

For the membrane stress ratio:

$$SR = \sigma_\alpha/\sigma_{HOOP} \quad \text{(Eq 9.9)}$$

where the stress ratio is an important design parameter to control pressure vessel failure modes:

$$\text{TPI} = \text{THICK}_{PLY} \cdot VF/\text{CSA} \quad \text{(Eq 9.10)}$$

where CSA is the tow cross-sectional area and can be calculated from yield (g/m) divided by the density (g/cm$^3$), TPI is the tows per inch or tows per millimeter, and $\text{THICK}_{PLY}$ is the total thickness of either the hoop or helical layers divided by the number of plies.

The dome thicknesses down to the inflection point can be calculated for designs where the band width, $BW$, is small compared to the radius of the pressure vessel. This thickness is increased for any local reinforcement added to the domes:

$$\text{THICK}_{DOME} = P \cdot R_{LOCAL} \cdot (1 - (R_{CYL}^2/R_{LOCAL}^2) \cdot \tau)/(2 \cdot \sigma_{DOME} \cdot VF \cdot \cos^2(\alpha) \cdot \sin(\varphi)) \quad \text{(Eq 9.11)}$$

where $\tau = \text{Thrust}/(P\pi R_{CYL}^2)$ can be used for dome shaping, $\alpha$ is the local wind angle, $\varphi$ is the local normal angle to the dome meridional shape (starts at zero at dome cylinder junction), and $R_{LOCAL}$ is the local radius where thickness is calculated.

This relationship can be simplified:

$$THICK_{DOME} = THICK_{HELICAL} \cdot \cos(\alpha_{CYL})/\cos(\alpha) \cdot R_{CYL}/R_{LOCAL} \quad \text{(Eq 9.12)}$$

when the dome is a continuation of the cylinder helical layers. This equation breaks down at the port, and overlapping ply-counting routines are added to correct for band width effects. The general equation is often used down to one band width away from the port radius as a first approximation, and more sophisticated approaches use the overlapping ply-counting routines for the whole dome (Ref 9.7). $\sigma_{DOME}$ is the allowable dome helical fiber strength and is typically equal to $\sigma_{\alpha}$, unless the dome is reinforced with a partial helical layer covering only the dome.

For winding purposes, the TPI is used to control the actual thickness of fiber laid down for hoop windings, while the $BW$ is typically used to define the helical wind pattern along with the helical angle, $\alpha$. The effective hoop wind angle is near 90° but is more accurately defined by the pressure vessel diameter, $D$, and band width or band advance per revolution of the mandrel, which is $90 - \tan^{-1}(BW/D)$ degrees. Hoop plies are generally wound in pairs or layers so that the slight high-angle helical angle of the hoop ply is balanced:

$$BW = N_{TOWS}/TPI \quad \text{(Eq 9.13)}$$

where $BW$ is the band width of the grouped tows, and TPI is the number of tows per inch or tows per millimeter. (Note: 1/TPI = tow width, which is generally limited for prepregs.)

Practical band widths are 2 to 4% of the vessel diameter, or the number of tensioners available on the winding machine times the tow width, whichever is the lesser.

The integer number of tows can be verified by:

$$N_{TOWS} = T_{PLY} \cdot BW/CSA \quad \text{(Eq 9.14)}$$

For helical tows:

$$N_{HELICAL} = THICK_{HELICAL} \cdot VF_{HELICAL} \cdot BW_{HELICAL}/CSA/N_{PLYS} \quad \text{(Eq 9.15)}$$

where $BW$ is band width, and $N_{TOWS}$ is the number of tows (spools) per band. To begin, approximate by using $THICK_{HELICAL}$/0.010 in. (0.254 mm) for the helical plies and $THICK_{HOOP}$/0.010 in. (0.254 mm) for the hoop plies.

For hoop tows:

$$N_{HOOP} = THICK_{HOOP} \cdot BW_{HOOP} \cdot VF_{HOOP}/CSA/N_{PLIES} \quad \text{(Eq 9.16)}$$

**Numerical Example for a Cylinder.** The netting analysis equations are used to determine wall thickness for a 12.0 in. (304.8 mm) inside diameter pressure vessel (RMC) of approximately 20.2 in. (513.08 mm) in length designed to contain a 2025 in.$^3$ (33,184 cm$^3$) volume. From overall vessel geometry and the required opening sizes (Fig. 9.5, a derivative of ASTM D 2585, Ref 6), the cylinder wind angle is determined to be 26.5°. The vessel is to operate at 2795 psi (19,271 kPa) and have a 1.25 factor of safety at burst. High-strength, intermediate-modulus graphite roving M30SC in an 18K tow (specific gravity 1.73 g/cm$^3$; tensile strength 800 ksi, or 5516 MPa; and modulus 42.7 × 10$^6$ psi, or 294 GPa) and a CSA of 6.8122 × 10$^{-4}$ in.$^2$ (0.4393 mm$^2$) is selected for winding. The design

**Fig. 9.5** Typical pressure vessel

allowables are 560 ksi (3843 MPa) helical/polar fiber stress and 740 ksi (5123 MPa) hoop fiber stress. The selected resin is an amine-cured wet winding resin with a specific gravity of 1.22 g/cm$^3$. The geometry of the vessel indicates the length-to-diameter ratio ($L/D$) to be 1.65, and the vessel is helically wound with a helical band width of 0.459 in. (11.66 mm). A 26.5° (note: the geodesic angle is 32.3°) leading sequential helical pattern is used, which closes in 74 circuits (number of winding bands to cover the entire cylindrical surface). The wind pattern is an $M$ = 1 (mandrel rotation), $N$ = 3 (star pattern), with $\alpha_T$ = 121.6° (mandrel rotation per circuit), and the $\beta/\theta = 25$ to ensure closure. The hoop band width is 0.434 in. (11.02 mm). (The number of tows per band is four for both helical and hoop layers. The lay-up consists of two hoop and two helical layers fully interspersed ($OXOX$, where $O$ is the hoop layer and $X$ is the helical layer). Note that a typical ply thickness ($T$) is approximately $T_{REF} \cdot CSA/CSA_{REF}$ (0.0056 in. × 6.8122/3.815, or 0.142 mm × 0.4393/0.2461), which equals 0.010 in. (0.254 mm) (including resin for this large tow fiber. The thrust and moment for this application are set to zero.

The netting analysis will define the required fiber thickness in the hoop and helical or polar directions. With these known, the designer can select other key parameters. The following paragraphs determine fiber volume, winding band width (hoop and polar), number of layers, composite wall thickness, and number of spools (hoop and polar).

The fiber volume is determined by the fiber and resin weight percentages and the densities or specific gravities of the fiber and resin. For this example, the resin weight percentage is 32%:

$$\begin{aligned} VF &= 1/(1 + (W_R/(1 - W_R)) \cdot \rho_F/\rho_R) \\ &= 1/(1 + 32/68 \cdot 1.73/1.22) \\ &= 0.5998 \text{ or } 60\% \end{aligned} \qquad \text{(Eq 9.17)}$$

for both helical and hoop layers.

To calculate the required helical composite thickness, thrust and moment are set to zero for this example, as noted previously:

$$\begin{aligned} \varepsilon_{HOOP} &= 5123 \text{ MPa}/294{,}000 \text{ MPa} \\ &= 743{,}092 \text{ psi}/42.7 \times 10^6 \text{ psi} \\ &= 0.0174 \text{ mm/mm } (0.0174 \text{ in./in.}) \end{aligned} \qquad \text{(Eq 9.18)}$$

$$\begin{aligned} N_\varphi &= 1.25 \cdot 19{,}270{,}840 \text{ N/m}^2 \cdot 0.1524 \text{ m} \\ &\quad \cdot (1 + 0.0174)/2 \\ &= 1{,}867{,}486 \text{ N/m} \end{aligned} \qquad \text{(Eq 9.19)}$$

$$\begin{aligned} THICK_{HELICAL} &= (1{,}867{,}486 \text{ N/m})/0.60/ \\ &\quad \cos^2(26.5)/3842.58 \times 10^6 \text{ N/m}^2 \\ &= 1.01135 \text{ mm } (0.03982 \text{ in.}) \\ &\quad \text{(fiber and resin)} \end{aligned} \qquad \text{(Eq 9.20)}$$

$$\begin{aligned} N_{PLIES} &= 1.01135/0.254 \\ &= 3.98 \text{ or four plies (two helical layers)} \end{aligned} \qquad \text{(Eq 9.21)}$$

With two layers or four plies selected, the actual helical ply thickness becomes 0.009954 in. (0.25284 mm).

The required hoop fiber thickness is:

$$\begin{aligned} N_{HOOP} &= 1.25 \cdot 19{,}270{,}840 \text{ N/m}^2 \cdot 0.1524 \text{ m} \\ &\quad \cdot (1 + 0.0174) \\ &= 3{,}734{,}972 \text{ N/m} \end{aligned} \qquad \text{(Eq 9.22)}$$

$$\begin{aligned} THICK_{HOOP} &= (3{,}734{,}972 - 1{,}867{,}486 \\ &\quad \cdot \tan^2(26.5))/0.60/5{,}123.437 \times 10^6 \\ &= 1.06400 \text{ mm } (0.04189 \text{ in.}) \\ &\quad \text{(fiber and resin)} \end{aligned} \qquad \text{(Eq 9.23)}$$

$$\begin{aligned} N_{PLIES} &= 1.06400/0.254 \\ &= 4.189 \text{ or four plies (two hoop layers)} \end{aligned} \qquad \text{(Eq 9.24)}$$

With four plies selected, the actual hoop ply thickness becomes 0.010472 in. (0.266 mm).

Total composite thickness becomes:

$$THICK = 1.064 + 1.01135 = 2.07535 \text{ mm } (0.082 \text{ in.}) \qquad \text{(Eq 9.25)}$$

The stress ratio becomes:

$$SR = 557{,}319/743{,}092 = 0.75 \qquad \text{(Eq 9.26)}$$

Using this information, wind parameters such as the number of roving spools required and band width can be derived. The typical winding band widths that are used for helical winding are 2 to 4% of the vessel diameter or controlled by the winding equipment available and the number of tensioners available.

For helical or polar windings, the wider the band the better, up to the point where the material becomes overstressed at the port. (However, a good rule of thumb is to limit the maximum band width to approximately 4% of the diameter.) For hoop winding, the band width is generally selected to use the same number of spools or tows as the helical or polar windings, so that the fiber delivery system is not modified when changing from hoop to helical winding and vice versa.

To complete the calculation, the designer must know the cross-sectional area of the fiber selected. In this case, the cross-sectional area of M30SC-18K graphite roving is $6.8122 \times 10^{-4}$ in.$^2$ (0.4393 mm$^2$), as shown in Table 9.3. The fiber lay-down or tows per inch (TPI), band width ($BW$), and number of tows ($N_{TOWS}$) are calculated as follows.

For helical tows:

$$TPI_{HELICAL} = T_{PLY}/CSA = 0.25284 \cdot 0.60/0.4393 = 0.3453 \text{ tows/mm } (8.77 \text{ tows/in.}) \quad \text{(Eq 9.27)}$$

$$\text{Estimated } BW = \text{Outside diameter} \cdot (0.02 \text{ or } 0.04) = (308.97 \text{ mm}) \cdot (0.02 \text{ or } 0.04) = 6.1 \text{ to } 12.45 \text{ mm } (0.24 \text{ to } 0.49 \text{ in.}) \quad \text{(Eq 9.28)}$$

$$N_{TOWS} = T_{PLY} \cdot BW/CSA = (0.25284 \cdot 0.60) \cdot 6.1/0.4393 = 2.11 \text{ or two tows to } 0.25284 \cdot 0.60 \cdot 12.45/0.4393 = 4.3 \text{ or four tows} \quad \text{(Eq 9.29)}$$

Choose the larger of the two choices, or four tows, to minimize the helical wind times and circuits to finish each layer. The actual $BW$ becomes:

$$4/TPI = 4/0.3453 \text{ or } 11.584 \text{ mm } (0.456 \text{ in.}) \quad \text{(Eq 9.30)}$$

The actual winding pattern selected used a $BW =$ 0.459 in. (11.659 mm), which is within 0.66%.

For hoop tows:

$$TPI_{HOOP} = 0.26599 \cdot 0.60/0.4393 = 0.3633 \text{ tows/mm } (9.23 \text{ tows/in.}) \quad \text{(Eq 9.31)}$$

The hoop band width, $BW$, is four spools wound at 0.3633 tows/mm (9.23 tows/in.), or:

$$BW = N_{TOW}/TPI = 4/0.3633 = 11.010 \text{ mm } (0.433 \text{ in.}) \quad \text{(Eq 9.32)}$$

A quick check of the numerical example is worthwhile to validate the thicknesses and the stresses at pressure. If the following values are given:

- Inside radius = 6.00 in. (152.40 mm)
- Wind angle, $\alpha = 26.5°$
- Proof pressure = 2795 psi (19,270.85 kPa)
- Safety factor = 1.25
- Burst pressure = 3494 psi (24,090.28 kPa)
- Fiber volume for both helical and hoop plies = 60%
- Helical ply thickness = 0.009954 in. (0.2528 mm) per ply
- Helical TPI = 8.7672 tows/in. (0.3453 tows/mm)
- Number of helical plies = 4
- Helical layers = 2
- Hoop ply thickness = 0.010472 in. (0.266 mm) per ply
- Hoop TPI = 9.2235 tows/in. (0.3633 tows/mm)
- Number of hoop plies = 4

Then the total thicknesses of reinforcement are given by:

- Helical thickness = 0.2528 · 4 = 1.0113 mm (0.03982 in.)
- Hoop thickness = 0.266 · 4 = 1.0640 mm (0.04189 in.)
- Total thickness = 2.07535 mm (0.082 in.)

To calculate fiber stress, using Eq 9.3, 9.6, and 9.7 and rearranging terms:

$$\sigma_{HOOP} = P \cdot R_{CYL} \cdot (1 - ½ \cdot \tan^2(\alpha_{CYL})) /(THICK - P \cdot R_{CYL}/E_{FIBER} + P \cdot R_{CYL}/2/E_{FIBER} \cdot \tan^2(\alpha_{CYL})) = 743{,}109 \text{ psi } (5{,}123.556 \text{ MPa}) \text{ with hoop strain of } 1.74\%$$

Using Eq 9.3 and 9.4 and calculating hoop strain from rearranged Eq 9.3, 9.6, and 9.7 gives:

$$\sigma_{HELICAL} = P \cdot R_{CYL} \cdot (1 + \sigma_{HOOP}/E_{FIBER}) /THICK_{HELICAL}/\cos^2(\alpha_{CYL}) = 557{,}334 \text{ psi } (3{,}842.683 \text{ MPa}) \text{ with hoop strain of } 1.74\%$$

The desired design stresses (557,319 psi, or 3,842.579 MPa, helical and 743,092 psi, or 5,123.439 MPa, hoop) were exactly matched (within round-off error) using the netting analysis tools.

### *Dome Contours*

The dome contours used on RMCs and pressure vessels are predominantly geodesic or modified geodesic domes (also called Nourse domes or balanced ovaloid domes). The shape is derived from the following conditions:

- The filament stress is the same at every point on the dome (for true geodesic shapes) or varies along the dome (modified geodesic).

- The principal loads in the surface of revolution are reacted in the direction of the filament.

The contour itself is derived by an iterative process using numerical methods for solving the equations discussed as follows (Ref 9.8, 9.9). The derivation of these equations has been solved by several investigators. These contours are used today for the starting geometries and modified by finite-element analysis as required to achieve optimum results for a particular design.

The derivation shows that the only geometric variables that affect the dome shape are the mean radius of the cylinder, $R$, and the port radius, $R_E = R_P + BW/2$, to the center of the winding band at the polar opening. The thicknesses of the helicals (or polars) at the tangent line do not affect the calculation of the dome centerline, but they are used to define the inside dome coordinates. The theoretical shape of a filament-wound dome is shown in Fig. 9.6. The equilibrium equations used to calculate the contour do not consider the polar boss geometry. When the wind angle, $\alpha_{LOCAL}$ reaches a value of 54.74° ($\tan^2(\alpha_{LOCAL}) = 2$ and $\tau$ is 0), the meridional radius of curvature, $R_M$, must change sign (or reverse curvature occurs). This point on the dome is called the inflection point, $R_C$; to avoid reverse curvature, the iterative process is discontinued. Often (but not necessarily), the outer tip of the polar boss is placed at this point and the contour continued by blending into a cone, spherical, or elliptical shape. As the filament leaves the dome and continues over the polar boss, the wind angle, $\alpha$, increases to 90° as the fiber becomes tangent to the polar opening.

The general solution, by numerical integration, of the following equations from the dome cylinder junction (tangent line) down to the port radius is the basis for almost all dome contours:

$$\mathrm{Abs}(D\varphi/\mathrm{DARCLENGTH}) = (2/a - (E_{22}/E_{11} + \tan^2(\alpha))/(E_{22}/E_{11} \cdot \tan^2(\alpha) + 1)) \cdot \sin(\varphi) / R_{LOCAL} \quad \text{(Eq 9.33)}$$

where $a = (1 - (R_{CYL}/R_{LOCAL})^2 \cdot \tau)$, $E_{22}$ is the transverse helical ply modulus (zero for netting or reduced $E_{22}$ for crazed structure, or transverse cracking), and $E_{11}$ is the longitudinal helical ply modulus. $\tau$ is the dimensionless shaping factor, $\tau = \mathrm{Thrust}/P/\pi/R_{CYL}^2$. Most derivations exclude the $\tau$ factor; however, for aerospace application, it is often used to shape the dome along with the choice of starting $\alpha$. Typically, this equation is applied to the middle of the dome thickness and inside contour derived by taking ½ the local thickness normal to the contour. This offset transitions to a constant down at the port to avoid the shell normal from exceeding 90°:

$$\text{Local } \alpha \text{ (empirical)} \approx \sin^{-1}(R_{PORT}/R_{LOCAL}) + (\alpha_{CYL} - (\alpha_{AFT} \text{ or } \alpha_{FWD})) \cdot ((R_{LOCAL} - R_{PORT})/(R_{CYL} - R_{PORT}))^N \quad \text{(Eq 9.34)}$$

where N = ½ for most applications.

$R_{PORT}$ is the local radius at the inside diameter of the dome windings at the center of the winding band. The local $\alpha$ (previously defined) can be better defined taking into account friction on the dome (a much more complicated numerical solution—three nonlinear partial differential equations).

This equation is solved down to the inflection point ($2/a = \tan^2(\alpha)$) when the local $\alpha$ equals 54.74° and the contour is extended to the port radius, either by a spherical radius or a cone tangent to the local contour at the inflection.

For more sophisticated dome solutions that take into account the local fiber friction on the dome, the following equations were derived from Ref 9.10.

The general friction model governing equations are a function of the geometry radii and curvatures as well as the local surface friction between the fiber and the dome surface generated by the winding tension (Ref 9.10):

$$Dr/Dz = u \text{ (local slope)} \quad \text{(Eq 9.35)}$$

$$Du/Dz = ((1 + u^2)/r) \cdot (\tan^2(\alpha) - 2/a) \text{ (rate of change of the local slope)} \quad \text{(Eq 9.36)}$$

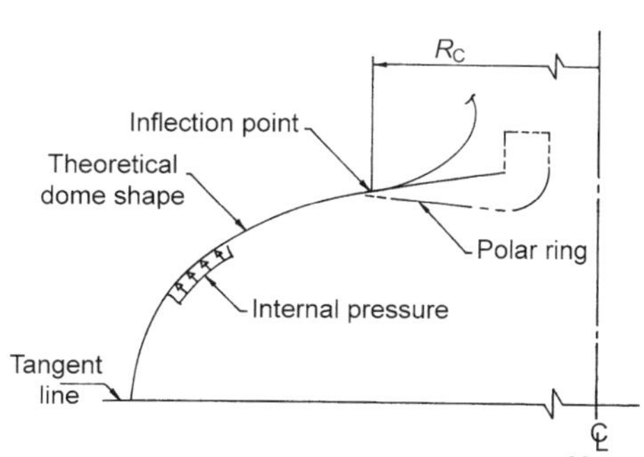


**Fig. 9.6** Typical dome contour geometry

$D\alpha/Dz = k \cdot (u \cdot \cos(\alpha)/(1 + u^2) - \tan(\alpha)/r) - u \cdot \tan(\alpha)/r$ (friction effect) (Eq 9.37)

where $r$ is the local radius as a function of axial position $z$, $\alpha$ is the local wind angle at $r$, k is the local friction coefficient (0.10 for carbon fiber/epoxy), and $a = 1 - (r_C/r)^2 \cdot \tau$ (where $r_C$ is the tangent line radius, and $\tau$ is the dimensionless thrust load on the dome).

These three equations are solved simultaneously down to the inflection point (local wind angle = 54.74°, as defined previously) using advanced numerical methods (Runge Kutta). The solution defines the local $r(z)$ and local $\alpha$ for a given k, given a starting cylinder radius and port radius. The k is adjusted to obtain a usable solution from the cylinder-to-dome junction down to the inflection point and as long as the friction capability is greater than the friction assumed for the contour solution. Otherwise, if the friction capability is less than the friction assumed, the fiber will slide either off the dome (negative deviation from geodesic) or toward the port (positive deviation from geodesic), causing bunching around the polar boss. Existing domes can be analyzed using these equations to see if stable winding can be achieved by solving for k and $\alpha$. It should also be noted that the special solution when k = 0 is the expected geodesic solution.

A special case solution of Eq 9.35 through 9.37 is for winding on a cylinder. $\alpha_{CYL}$ may be varied in the cylinder if sufficient cylinder length is available:

$$1/(\sin(\alpha_{AFT})) - 1/(\sin(\alpha_{FWD})) = k \cdot L_{CYL}/R_{CYL} \quad \text{(Eq 9.38)}$$

where k is the cylinder fiber path friction coefficient ≈ 0.10 for wet impregnated carbon fibers. The geodesic path in a cylinder is a constant wind angle.

Table 9.7 shows the friction coefficients that can be expected for typical fiber-resin combinations. Actual capability should be verified for the actual fiber-resin combination used.

The solution from the inflection to the port radius is either a spherical or conical closure surface with matching slope at the inflection, and the wind angle changes from 54.74° ($\tau$ = 0) to 90° using a geodesic formulation (i.e., $\alpha = \sin^{-1}(R_{PORTMIDBAND}/R_{CYL})$).

Corresponding mandrel rotation ($\theta$, also known as $\alpha_T$) for one complete circuit is the integral of $D\theta/Dz$:

$$D\theta/Dz = (1 + Dr/Dz^2)^{1/2} \cdot \tan(\alpha)/r \quad \text{(Eq 9.39)}$$

**Table 9.7 Typical friction coefficients for fiber-resin combinations**

- Glass/epoxy
  Wet: k = 0–0.18
  Prepreg: k = 0–0.35
- Aramid/epoxy
  Wet: k = 0–0.14
  Prepreg: k = 0–0.30
- Carbon/epoxy
  Wet: k = 0–0.11
  Prepreg: k = 0–0.20

Friction coefficients are determined empirically.

Examples of solutions to Eq 9.33 and Eq 9.35 through 9.37 are shown in Fig. 9.7 for several representative starting angles compared to a geodesic dome. As noted in the figure, the selected dome shape for this application was a 5° flattened, balanced Nourse or modified geodesic ovaloid. The flattening of the dome is normally used by aerospace companies to maximize pressure vessel volume for fixed-length systems. This technique also increases the dome axial deformation under pressure—a desired characteristic to aid in propellant loading and subsequent pressure cure. The downside of this type of dome shape is an increase in helical fiber stress at the port, which makes it susceptible to buckling. Therefore, detailed finite-element analysis should be used to determine how much flattening can be tolerated.

**Simplified Design Relationships for Domes.** To perform preliminary design or estimate material requirements, it is necessary to solve the dome contour equations. Figures 9.8 and 9.9 can be used to estimate the weight of the domes and the enclosed volumes (Ref 9.5). The inputs include:

- $R_o$ is the radius to center of the winding band adjacent to the boss, also known as $R_E$.
- $R_i$ is the radius of the cylinder (midthickness), also known as $R_{CYL}$.
- $T$ is the thickness of the helical or polar layers, also known as $THICK_{HELICAL}$.
- $\rho$ is the density of the composite, also known as $\rho_{HELICAL}$.

Note that the weight calculations can be performed using either composite or fiber densities to solve for the applicable weights.

These equations are applicable to either RMCs or pressure vessels using the geodesic dome

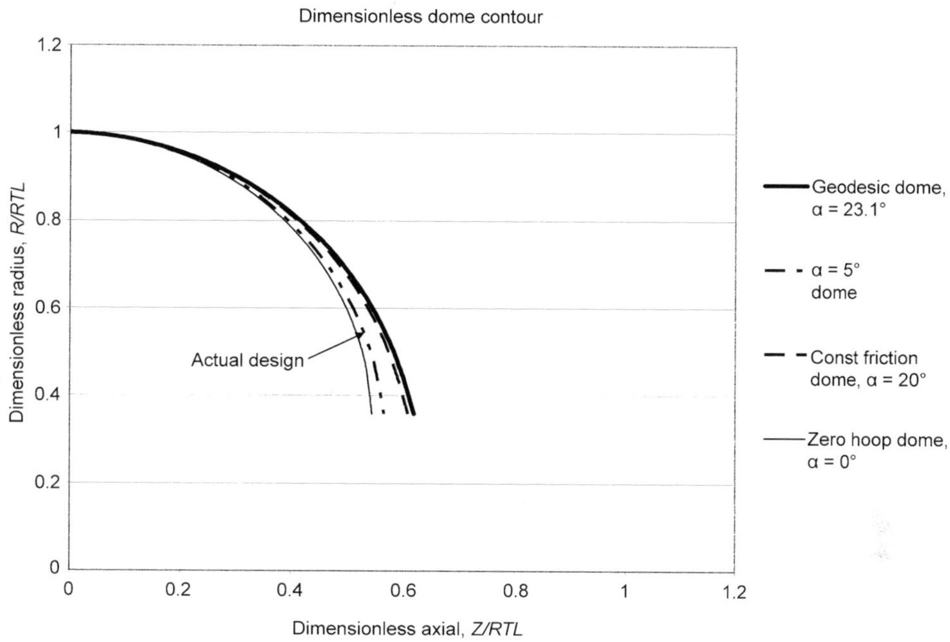


**Fig. 9.7** Typical dome contours

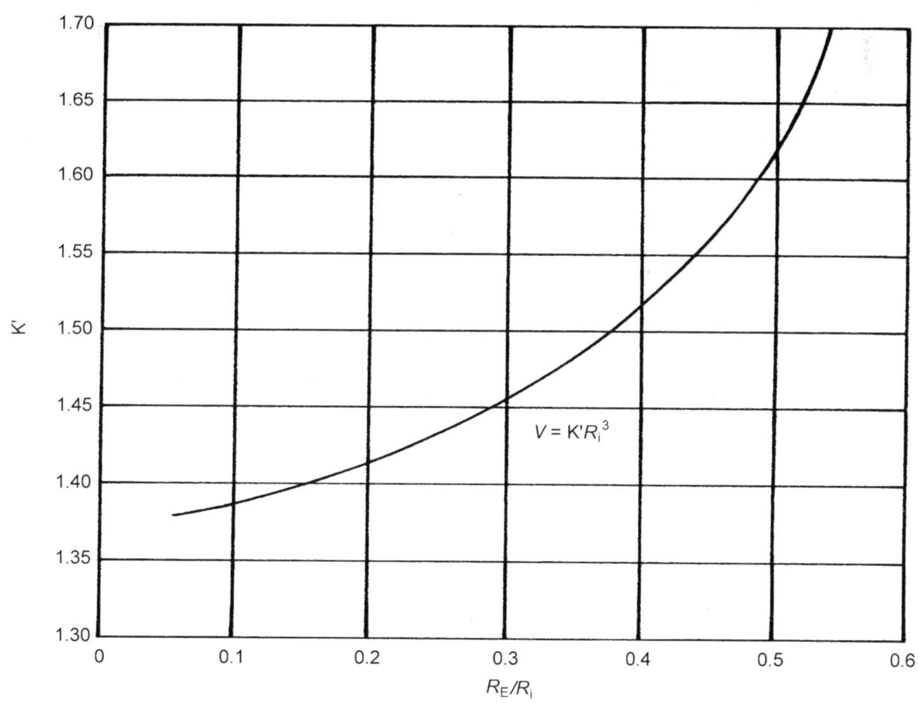


**Fig. 9.8** Geodesic dome volume nomogram

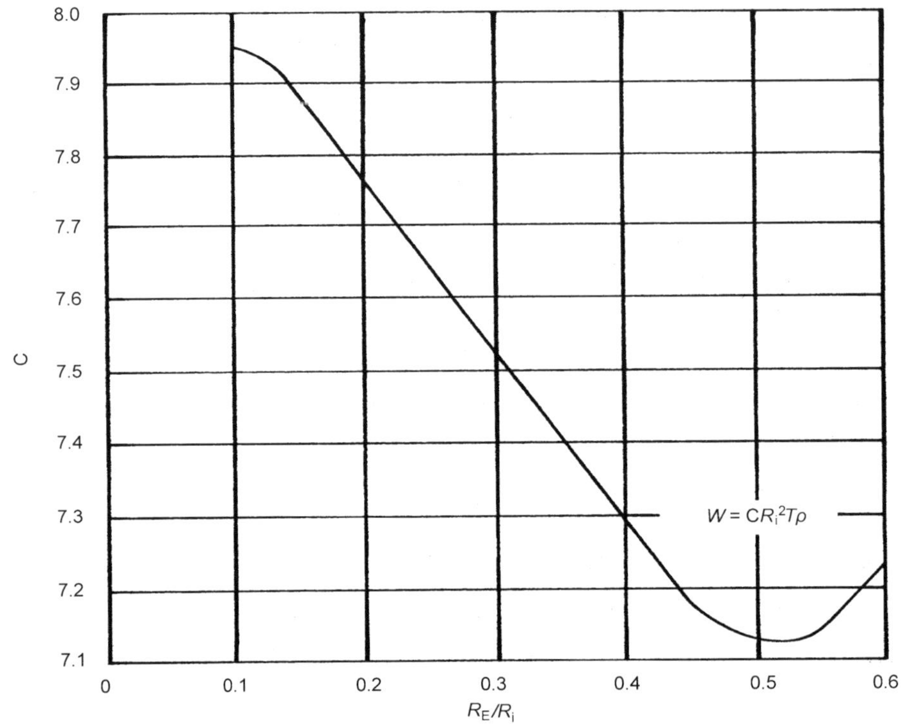


**Fig. 9.9** Weight of geodesic dome nomogram

shape. Exact final contours, of course, require computer solutions that can compensate for the deflected dome shape. Also, computer solutions include options that account for the RMC thrust component (based on the nozzle opening and internal pressure). Nonetheless, Fig. 9.8 and 9.9 give close approximations and are adequate for quick estimates and tool layouts. The total volumes and weights can be calculated:

$$V_{TOT} \approx \pi \cdot R_{CYL}^2 \cdot (L_{CYL} + 2/3 \cdot (H_{DFWD} + H_{DAFT}) \quad \text{(Eq 9.40)}$$

where $V_{TOT}$ is the total pressure vessel volume, $R_{CYL}$ is the cylinder radius, $L_{CYL}$ is the cylinder length, $H_{DFWD}$ is the forward dome height, and $H_{DAFT}$ is the aft dome height, which can also be expressed in terms of Fig. 9.7:

$$\begin{aligned} W_{TOT} \approx\ & 2 \cdot \pi \cdot R_{CYL} \cdot (L_{CYL} \cdot (\rho_{HELICAL} \\ & \cdot \text{THICK}_{HELICAL} + \rho_{HOOP} \cdot \text{THICK}_{HOOP}) \\ & + (\rho_{SKIRTAFT} \cdot LS_{AFT} \cdot T_{SKIRTAFT} \\ & + \rho_{SKIRTFWD} \cdot LS_{FWD} \cdot T_{SKIRTFWD})) \\ & + W_{DOMEAFT} + W_{DOMEFWD} + W_{BOSSAFT} \\ & + W_{BOSSFWD} + W_{YJOINTFWD} \\ & + W_{YJOINTAFT} + W_{FLANGEFWD} \\ & + W_{FLANGEAFT} + W_{FASTENERS} \quad \text{(Eq 9.41)} \end{aligned}$$

where $\rho$ is the density of the composite, $LS$ is the skirt length, and $W$ is the weight of the component. Dome weights ($W_{DOMEAFT} + W_{DOMEFWD}$) are estimated from Fig. 9.8 using the $R_{CYL}$ and $R_E/R_{CYL}$ ratio (port percentage) for each dome.

The pressure vessel efficiency ($PV/W$) for the pressure shell or the total case can be estimated using the aforementioned relationships.

## *Design Approach*

In addition to pressure-induced design considerations, there are the external loads, stiffness, need for local reinforcements, skirt and skirt-to-case joint design, polar boss and associated closure attachment design, and attach flange design, including bolted and riveted joints.

**Stiffness** as a design constraint or requirement is an important consideration. The different types include bending stiffness ($EI$, where $E$ is the axial modulus, and $I$ is $\pi \cdot R_{CYL}^3 \cdot T$), when the structure acts like a beam; axial stiffness ($EA$, where $A$ is $2 \cdot \pi \cdot R_{CYL} \cdot T$), when the structure acts like a spring; and shear stiffness ($KAG$, $GJ$, where K = ½, $G$ is the in-plane shear modulus, and $J$ is $A \cdot R_{CYL}^2$), when the structure acts like a beam or torsion spring element, respectively. Sometimes, the stiffness requirements drive the design and must be evaluated, typically using the laminated plate theory in Ref 9.11. The reductions in stiffness for the cylindrical section (including skirt/Y-joint) that are caused by crazing (transverse modulus degradation due to matrix, or resin, cracking) must be included.

**Discontinuities.** To deal with sudden ply drops or layer terminations, such as partial helical layer termination in the cylindrical section or skirt or pressure vessel layer terminations, the Timoshenko simplified discontinuity influence parameter, β, has been used successfully to determine how far to extend cut layers or ply terminations when dealing with a termination discontinuity. This methodology lets the designer estimate (with good accuracy) ply termination locations away from a known discontinuity, such as the bending discontinuity associated with the junction of a cylinder and a dome or a transition from a very thick laminate to a thin laminate. The calculation estimates the distance needed to reduce the effect of the discontinuity to zero (insignificance). $\beta \cdot L$ should be between 3 and 5, although some designs have been able to get by with approximately 1.5, where $L$ is the length to damp out the discontinuity:

$$\beta = (A_{22}/D_{11}/4/R^2)^{1/4} \text{ and } 3 < \beta \cdot L < 5 \qquad \text{(Eq 9.42)}$$

where $A_{22}$ is the circumferential stiffness, $D_{11}$ is the axial bending stiffness, and $R$ is the radius to the discontinuity region (i.e., tangent line).

This approach has been successfully used for partial helical layers (Peacekeeper, Trident II D5, NAVYTWD, and many independent research and development pressure vessels) and aft-end full-diameter joint designs such as Thaad.

**Stress ratio** is the ratio of helical to hoop fiber stress in the cylindrical membrane portion of the pressure vessel. Figure 9.10 shows typical delivered hoop fiber strength (Ref 9.12).

Part of the design process is to select a stress ratio that will produce the desired failure mode and the margin of safety sought while minimizing manufacturing variability. This ratio is simply the helical (or polar) fiber stress divided by the hoop fiber stress.

Figure 9.10 plots the hoop fiber stress achieved versus the stress ratio used. Note that for ratios less than 90%, the hoop failure mode prevails, and for 100% and greater, a helical failure is almost guaranteed. In most cases, a hoop failure is preferred, because these failures are primarily affected by membrane loading, which is more predictable with less scatter in strength. Helical failures (and this includes all dome failures)

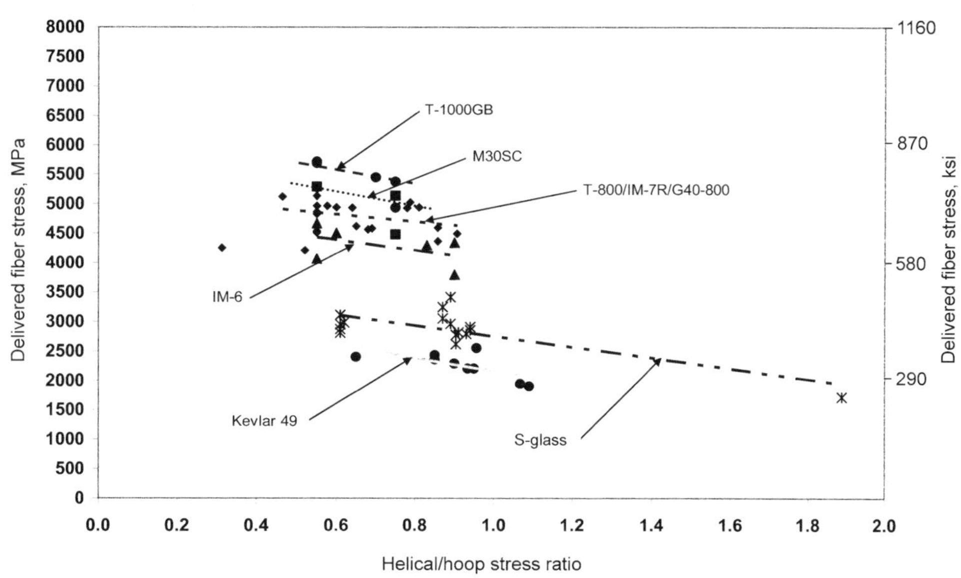


**Fig. 9.10** Industry-demonstrated average hoop fiber stress versus stress ratio (several rocket motor manufacturers)

are strongly affected by the bending loads present from the discontinuities at the tangent line and at the polar boss regions. For this reason, design allowable stresses for the helical fibers are almost always selected appreciably lower than hoop fiber allowables. Mixed-mode failures (stress ratios between 0.90 and 1.0), where hoop or helical failure occurs randomly, usually result in greater variability and are not recommended.

A comfortable stress ratio often used in RMC designs is 60 to 85% (normally no greater than 90%), which can further be assisted by using local dome reinforcement or even using an entire dome cap (partial helical layer).

The stress value selected for design will always be lower than that listed in the fiber suppliers' specifications. Exact values will vary between manufacturers; however, all values are discounted to some degree based on experience and to compensate for scatter. This reduction is known as the translation factor. This value may be anywhere from 70 to 95% (typically 85 to 95%) of tensile values specified by the fiber supplier. This value must be further reduced to allow for manufacturing variability (typically 6 to 12%).The higher value is associated with higher stress ratios and a variable failure mode.

### *Other Design Aspects*

**Skirt design** is driven by the need to efficiently carry compression and tension loads from the vehicle to the pressure vessel without introducing excessive stress concentrations. The structure must be optimized to reduce weight and complexity and to be compatible with geometric constraints. Several analytical approaches are used. All must derive a ply-stacking sequence that meets strength requirements while achieving buckling resistance at minimum weight constrained by manufacturing considerations. Typically, a quasi-isotropic laminate is chosen, although the optimum laminate consists of axial reinforcement (usually unidirectional), ±45° reinforcements (either unidirectional or woven), and circumferential windings in an optimum sequence. Ply stack optimization (minimum weight lay-up) is typically conducted for buckling while meeting the ply strength requirements using:

- NASA SP 8007 (matrix operations for solving the buckling modes)
- Air Force Advanced Design Guide (codeable equations similar to NASA Special Publications)
- BOSOR (shell computer code for performing buckling analysis)

Composite stress analysis, typically done by classical laminated plate analysis, is available in various forms:

- NASA Integrated Composite Analyzer (ICAN) laminate program
- Cylindrical laminate analysis program (CYLAN)

**Skirt Y-Joint Design.** The skirt-to-case Y-joint shear ply design is critical to transferring the skirt loads into the pressure vessel without introducing excessive stress concentrations. Types of joints include:

- Rubber shear ply with a shear modulus of approximately 200 psi (1380 kPa) skirt reinforcements extending beyond the shear ply. These create severe stress discontinuities on the inboard edge of the shear ply.
- Hard attachment (prone to shear-induced delaminations)
- Intermediate-modulus film adhesive materials with a shear modulus of 50 to100 ksi (345 to 690 MPa)

The design analysis technique is typically a shear lag analysis (one-dimensional quasi-axis symmetric). This is solved with a Runge Kutta methodology in a computer program. With appropriate boundary conditions ($\sigma_C$ is defined at the beginning and end of the shear ply, and $\sigma_S$ is defined at the beginning and end of the shear ply), the following equation is solved for the axial stresses and shear stresses for the skirt, the pressure vessel under the skirt, and the shear ply:

$$Dt/dz \approx (\sigma_C/EC - \sigma_S/ES)/(T_{SP}/G_{SP} + T_S/1.5/G_S + T_C/1.333/G_C) \qquad \text{(Eq 9.43)}$$

where $t$ is the shear stress in the shear ply (SP) as a function of axial position ($z$), $\sigma$ is the axial stress in either the skirt (S) or the case (C) as a function of axial position, $E$ is the local effective axial modulus of the skirt (S) or the case (C), $T$ is the local thickness as a function of axial position for either the skirt (S), the case (C), or the shear ply (SP), and $G$ is the shear modulus of the shear ply (SP), the case (C), or the skirt

(S) as a function of axial position. Because this is a one-dimensional stress and strain balance, effective modulus must be used for the case and skirt to account for the real biaxial local stress/strain field.

A biaxial approximation that has been used effectively for the skirt is:

$E_{EFF} \approx E_{AXIAL}/(1 - 2 \cdot \nu_{XY})$

and for the case under the shear ply:

$E_{EFF} \approx E_{AXIAL}/(1 - \tan^2(\alpha) \cdot \nu_{XY})$

where $E_{EFF}$ is the effective modulus, $E$ is the local Young's modulus in the axial direction, $\nu_{XY}$ is the local axial-hoop Poisson's ratio, and $\alpha$ is the local wind angle. The final verification and tailoring is done by finite-element analysis.

**Polar Boss Design.** The typical types of polar bosses (various metal and composite structures) are listed as follows and shown in Fig. 9.11 (Ref 9.13):

- Axial bolted protruding or submerged (bearing ring) with or without inserts with matching closures (conical, concave, flat)
- Snap ring/threaded ring protruding or submerged (bearing ring) with matching closures (conical, concave, flat)
- Radial bolt protruding or submerged (bearing ring) with matching closures (conical, concave, flat)

Close-form strength-of-materials solutions are typically used with empirical rotation (stiffness) correlation factors, including composite bearing pressure matched to industry designs. Finite-element analysis confirms and tailors the final design. The design is iterated until matched to composite dome deformation to minimize helical fiber discontinuities, peak helical fiber stress, and fastener stresses. Analysis must include case, shear ply, boss, fastener, and closure to accurately predict behavior. Figure 9.12 shows a typical axial bolted boss with associated nomenclature:

$$P_{BEAR} \approx 2 \cdot R_O^2 \cdot P_{ULT}/(R_O^2 - R_F^2) \qquad \text{(Eq 9.44)}$$

where $P_{BEAR}$ is the bearing stress on the composite, $R_O$ is the outside radius, $R_F$ is the inside radius of the windings, and $P_{ULT}$ is the ultimate design pressure (ignoring thrust relief). Allowed

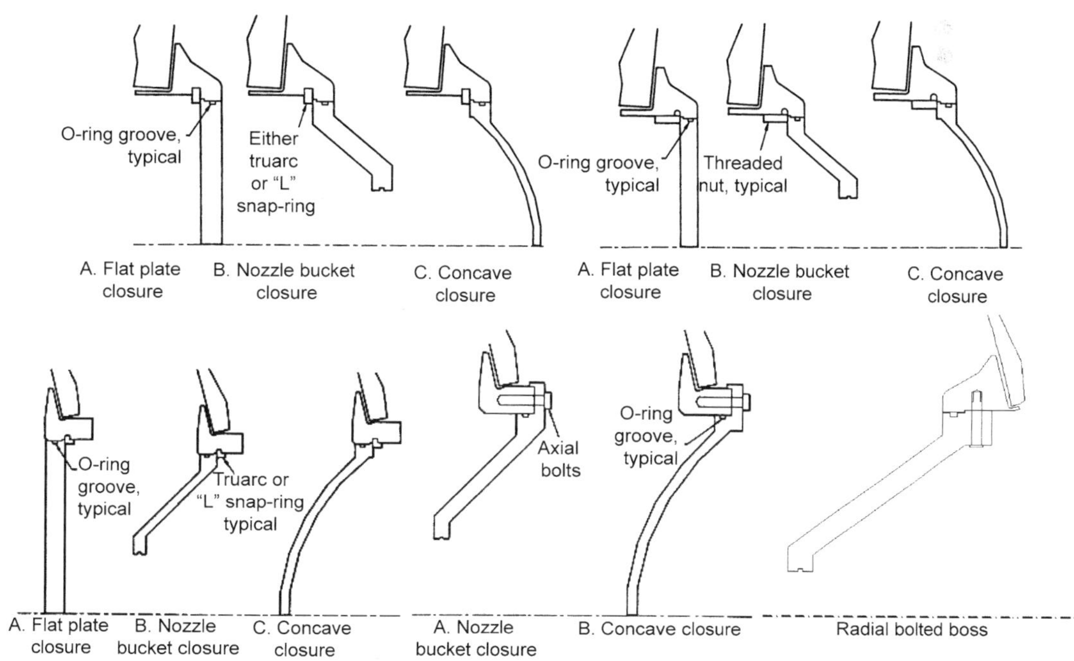


**Fig. 9.11** Polar boss and closure configurations

bearing pressure is a function of port radius/cylinder radius ($R_B/R_C$) and demonstrated $P_{BEAR}/P_{ULT}$ ratio. The maximum bearing pressure should be less than the laminate transverse compressive yield strength to avoid cracking. A good rule of thumb to keep the bearing pressure within demonstrated levels is to make the $R_O - R_F > 1.5 \cdot BW$ (band width):

$$T_F \approx ((R_O - R_F) \cdot R_O^2 \cdot P_{ULT}/(R_{NX} \cdot \sigma_{ULT}))^{1/2} \text{ and } T_F \leq 2 \cdot D_B \quad \text{(Eq 9.45)}$$

where $T_F$ is the boss flange thickness; $R_{NX}$, the radius to center of load, is $\approx R_F + 1/3 \cdot (R_O - R_F)$; $D_B$ is the bolt diameter; and $\sigma_{ULT}$ is the boss allowable stress. $R_F - R_B$ is usually set to $2 \cdot D_B$. Roark and Young ring and plate equations can also be used to estimate the flange thickness.

For axial bolted joints:

$$N_B = \pi \cdot R_B^2 \cdot P_{ULT} \cdot BF/F_{BULT} \text{ such that } 3 \leq 2 \cdot \pi \cdot (R_B + D_B)/N_B/D_B \leq 4 \quad \text{(Eq 9.46)}$$

where $N_B$ is the number of bolts, BF is the bending or prying factor, and $F_{BULT}$ is the bolt strength.

Rotation, $\theta$, is approximately:

$$\theta \approx M \cdot R_B^2/E_{BOSS}/I_{YY} \quad \text{(Eq 9.47)}$$

where $M \approx P_{ULT} \cdot R_B \cdot (R_{NX} - R_B)/2$; $R_{NX}$ is the radius as shown earlier; the allowed angle of rotation is a semilog function of port opening ratio $R_B/R_C$, where $R_C$ is the case cylinder radius; $I_{YY}$ is the rotational stiffness and is $\approx T^3 \cdot (R_F - R_B)/12$ plus corrections, where $T$ is the effective height of the boss, including closure flange thickness if applicable; and $E_{BOSS}$ is the boss Young's modulus.

For boss hoop stress:

$$\sigma_\theta \approx \theta \cdot X_{CG} \cdot E_{BOSS}/R_B \quad \text{(Eq 9.48)}$$

where $X_{CG}$ is the distance from the center of mass to the end of the boss or boss plus flange.

Snap ring, threaded ring, or radial bolts are sized by simple shear, simple bearing, and bending stresses using Roark and Young.

**Attach flanges** can be stand-alone rings or part of an interstage (Fig. 9.13). The attach flange ring is attached to the composite skirt using the following methods:

- Adhesive bonding—rarely used
- Rivet (or bolt)—common
- Adhesive and rivet (or bolted)—most common

The design analysis techniques used to size the flange for pin bearing, shear tear out, and net section tension are normally based on testing or prior experience. Joints are typically designed to be bearing-critical (Ref 9.13):

$$T_F \geq N_X \cdot 2 \cdot \pi \approx R/N_R/D_R/F_{BRUF} \quad \text{(Eq 9.49)}$$

or

$$T_F \geq N_X \cdot \pi \approx R/N_R/(1.5 \cdot D_R - 0.3535 \cdot D_R)/F_{SUF} \quad \text{(Eq 9.50)}$$

or

$$T_F \geq N_X/(1 - N_R \cdot D_R/2/\pi/R)/F_{TUF} \text{ (ignoring load eccentricities)} \quad \text{(Eq 9.51)}$$

where $T_F$ is the flange thickness on the case side or the interstage side, $N_X$ is the design axial line load (axial load/2/$\pi$/$R$), $R$ is the flange radius, $N_R$ is the number of rivets or bolts, $D_R$ is the diameter of the rivet or bolts, $F_{BRUF}$ is the flange ultimate bearing strength, $F_{SUF}$ is the flange shear strength, and $F_{TUF}$ is the flange tensile strength. Recommended edge distance ($e$) for rivets or bolts is $\geq 2.3 \cdot D_R$ (3.0 is a good starting point). Likewise, the distance circumferentially between fasteners, $W$, should be $\geq 2.76 \cdot D_R$, and $D_R$ should be between $0.78 \cdot T_{SKIRT}$ and $2 \cdot T_{SKIRT}$, where $T_{SKIRT}$ is the composite skirt thickness. Likewise, for aluminum the $e/D_R \geq 1.5$, $W/D_R \geq 3$, and $D_R/T_F$ should be between 0.78 and 2.0:

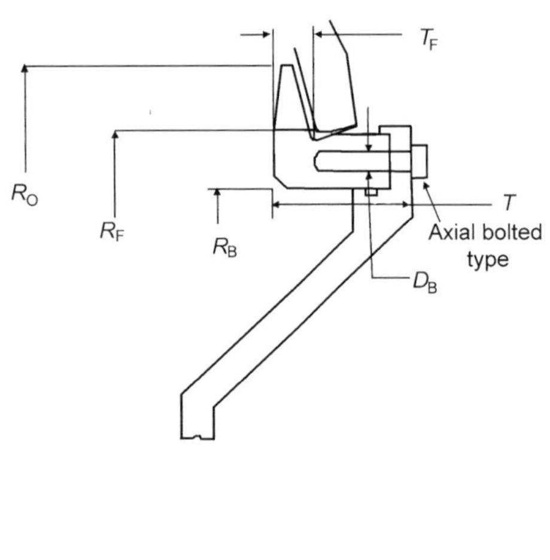


A. Nozzle bucket closure

**Fig. 9.12** Axial bolted boss geometry with closure

$$F_{TUF} \geq (N_X/T_F + 6 \cdot N_X \cdot \delta/T_F^2) \cdot (1/(1 - N_R \cdot D_R/2/\pi/R)) \quad \text{(Eq 9.52)}$$

where $\delta$ is the eccentricity between the attach flange at the composite skirt compared to the flange at the vehicle interface.

## Manufacturing

The manufacturing operations associated with RMCs are, by need and definition, repetitious. Quality and reliability must be maintained over hundreds of units over several years, thus requiring that each unit be identical to its counterpart. This requires precise control manufacturing and inspections. These objectives are controlled by using very detailed processing (travelers) and a dependable material control system.

To produce a finished product requires well-coordinated teams working under a rigid system. Figure 9.14 outlines the typical organizational structure required to produce an RMC. The person in control and at the center of all these disciplines is the program manager. Manufacture of the casing is just one of the critical items the program manager directs; however, this article

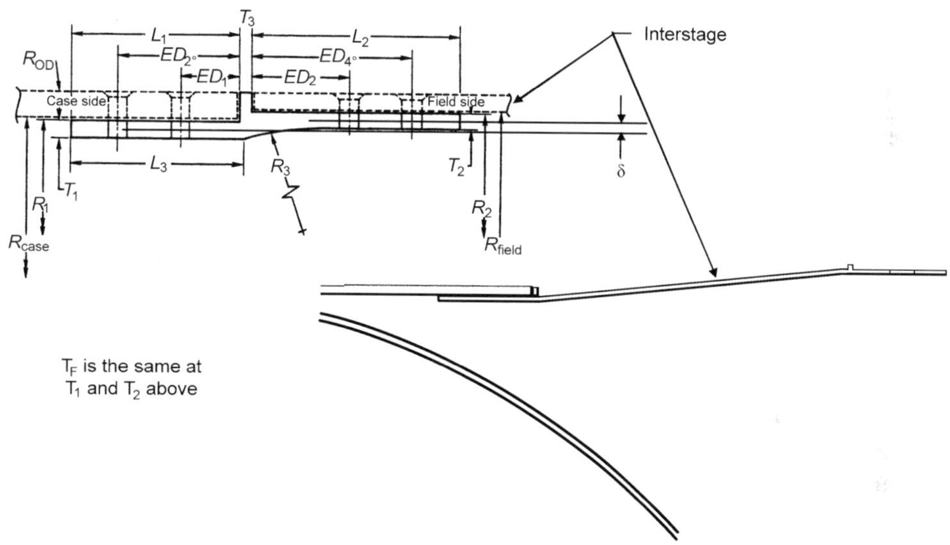


**Fig. 9.13** Attach flanges and interstage examples

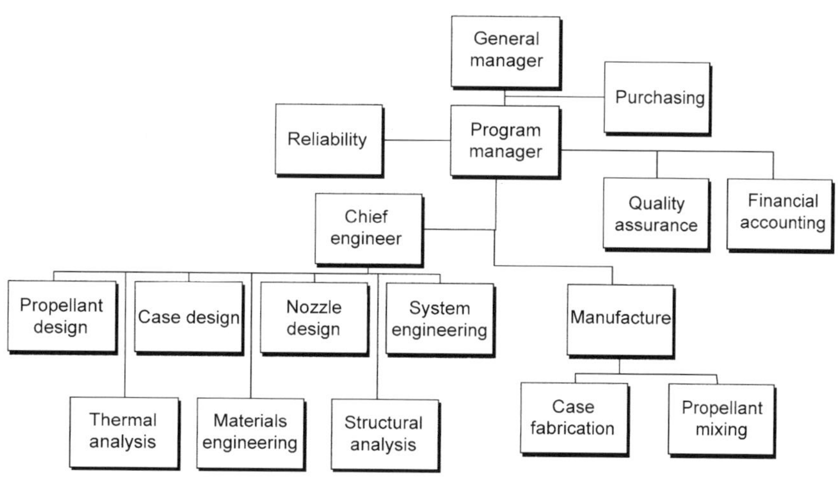


**Fig. 9.14** Typical rocket motor manufacturing organization

addresses only fabrication of the composite case. Figure 9.15 shows a manufacturing flow diagram outlining the key operations, which are described in the following paragraphs.

### Mandrel Selection

The key piece of tooling for manufacturing pressure vessels is the winding mandrel. The mandrel is the tool that provides the inside surface and shape for forming the wound product. Several mandrel types have been used for manufacturing: sand/polyvinyl alcohol (with or without lightning tubes), cast or swept plaster (with and without steel reinforcement), and segmented metal mandrels (steel, aluminum, and hybrids). In an effort to reduce tooling costs, several alternative mandrels have been evaluated. These include high-density foam, plaster with microballoons, pressurized plastic or metallic liners, and pressurized reinforced rubber. In the pressure vessels used for fuel tanks (CNG, fireman air tanks, etc.), it is common to use the liner (pressurized) as the mandrel. For rocket motor combustion chambers, it is common to use segmented metal mandrels. The insulator is almost always prefabricated and preassembled onto the mandrel along with the polar fitting and wind axis. Finally, the two insulator halves are spliced together using uncured cylinder rubber that will be co-cured with the case. An alternative approach for applying the insulation is to use a strip winding machine (wound elastomeric insulation) for the majority of the dome and cylinder using a lathe-type machine that extrudes the uncured rubber material to an exact thickness and width, which allows the insulation thickness to be tailored by location along the mandrel surface. Other versions of the uncured lay-up have also been used. Figure 9.16 shows several representative segmented metal mandrels.

For pressure vessels using metal or nonmetallic composite liners, as shown in the flow diagram in Fig. 9.15, an alternate technique is that of winding directly on the liner, which serves as the mandrel. This method is commonly used in pressure vessels, where the liner also serves as a permeation barrier. The liner is typically incrementally pressurized to offset the winding, cure induced loads, and prevent distortion.

The importance of the mandrel tooling is shown in Fig. 9.17. The winding loads are relatively small compared to the cure-induced loads of a segmented metal mandrel. The cure-induced loads often drive the design of the winding mandrel.

### Impregnation

The bringing together of resin and roving into a form suitable for winding has been discussed in Chapter 3 of this book or in Ref 9.5. The impregnation process demands consistency and uniformity of the impregnated product, whether it is wet wound, wet rerolled, or commercially prepregged.

### Mandrel Preparation

The assembled mandrel is prepared by cleaning, applying a release (if the mandrel is removed), and applying the insulator (rocket motor combustion chambers) or bonding adhesive for pressure vessels with liners.

### Trial Winding and Pattern Generation

Winding disks are prepared in accordance with the manufacturing engineer's direction,

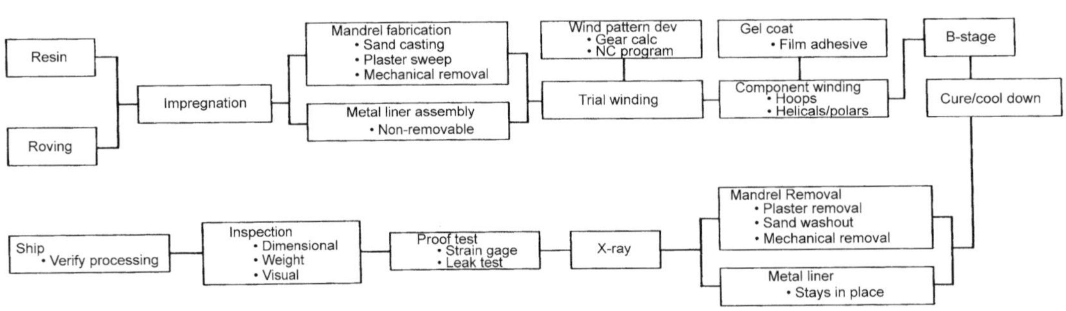


**Fig. 9.15** Rocket motor manufacturing flow diagram

consistent with the pressure vessel design documentation and approval of design engineering. When winding a prototype or revising a pattern, trial winding is mandatory. At this time, fine adjustments in the winding eye position, mandrel rotation, band width, and other parameters can be made. Minor changes to the delivery system and to the winding disk are finalized before going into production.

The theoretical total circuits to close for the $i$th layer, $C_{Ti}$, are:

$$C_{Ti} = \pi D_{MCMAXi} \cos(\alpha_{MAXi})/BW_i \qquad \text{(Eq 9.53)}$$

where $i$ is the $i$th helical layer, $BW_i$ is the band width for the $i$th layer, and $D_{MCMAXi}$ is the maximum cylinder diameter, where $\alpha_{MAXi}$ exists for the $i$th layer.

$C_T/N$ is an integer and is the theoretical circuits to close divided by the star pattern. $C_A = C_T \pm k$ is the actual total circuits-to-close pattern (completion of one helical layer). $\theta = 360/C_A$ is the angle in degrees that is related to maximum cylinder diameter, maximum $\alpha$ in the cylinder, and the band width.

Mandrel revolution for helical winding circuits (star pattern) is defined in terms of $M/N$,

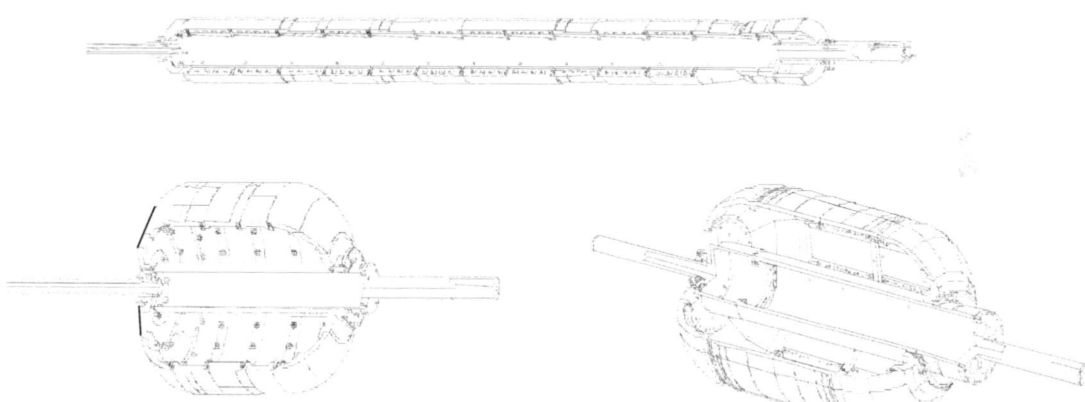

**Fig. 9.16** Typical segmented metal mandrels

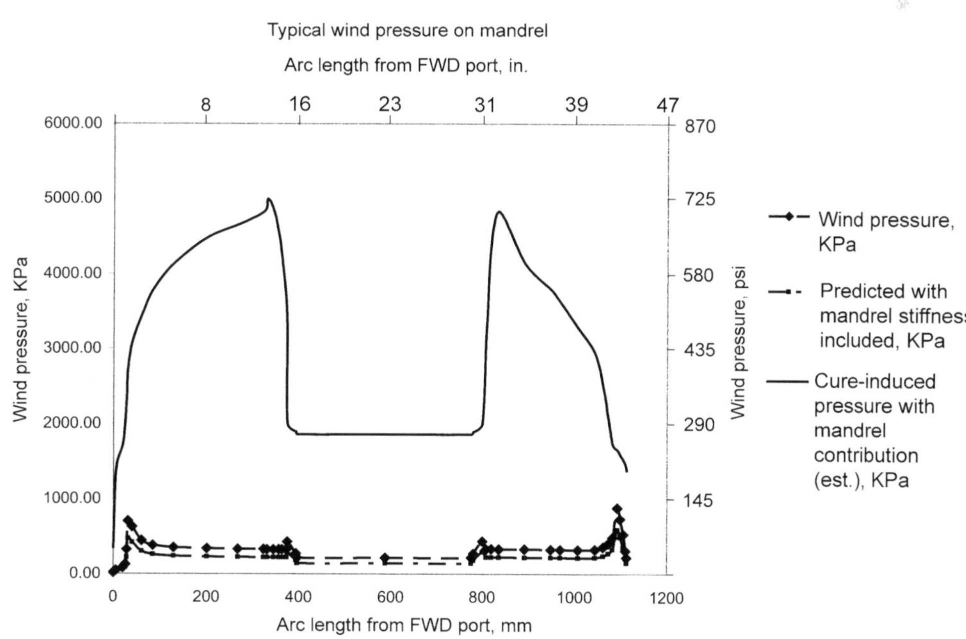


**Fig. 9.17** Mandrel loads during winding and cure

where $M$ is the mandrel rotations, and $N$ is the star pattern. The following relationships are used to finalize the winding pattern for each layer.

The total degrees of mandrel rotation per circuit are:

$$\alpha_T = 360\,(M/N) \pm (\theta/N) \qquad \text{(Eq 9.54)}$$

$\beta = (\alpha_T/360)$ – integer of $(\alpha_T/360)360$ and is the actual degrees of mandrel rotation between the tangent line exit point of any two consecutive circuits. $\beta/\theta$ is an integer to ensure closure will occur:

$$v = (MN) - (\text{Integer of } M/N) \qquad \text{(Eq 9.55)}$$

$$1/N = (ka) - (\text{Integer of } ka) \qquad \text{(Eq 9.56)}$$

where $k = 1$ results in a sequential pattern, $1 < k < N$ results in a nonsequential pattern, and $a = v$ if $v < 0.5$ or $a = (1 - v)$ if $v > 0.5$. In other words, $a = v$ or $a = 1 - v$. Typical winding patterns can be visualized as shown in Fig. 9.18.

### *Insulation or Liner Adhesive Application*

The entire insulation surface is grit blasted (or rough sanded), cleaned with a solvent, and a resin gel coat is applied to serve as an adhesive between the insulator and composite overwrap. The use of a film adhesive in place of the resin gel coat is becoming common. This provides a controlled adhesive thickness with repeatable properties; however, the cost can be a consideration.

### *Pressure Vessel Winding*

The RMC pressure vessel winding is closely controlled by the formal processing issued. Critical parameters are verified periodically by quality-control inspectors. The following minimum controls and suggested control methods will generally be imposed during development and less often during production:

- Roving width—micrometer
- Roving tension—tensiometer
- Roving thickness—on component with pi tape
- Resin viscosity—viscometer
- Resin-mix ratio verification—refractometer
- Resin temperature in bath—thermometer
- Resin content on roving—triple-beam balance
- Fiber lay-down angle—protractor

Any nonconforming items are "squawked" by the quality-control inspector. This action requires immediate disposition by the engineer. Actual examples of helical and hoop winding are shown in Fig. 9.19 and 9.20.

### *Skirt Fabrication*

The skirts are attached near the juncture of the domes and the cylinder and serve as the structure to convey loads between various stages (or the payload) of the missile (Fig. 9.21). These are attached by various bonding/winding or riveting methods to convey loads through the motor case assembly.

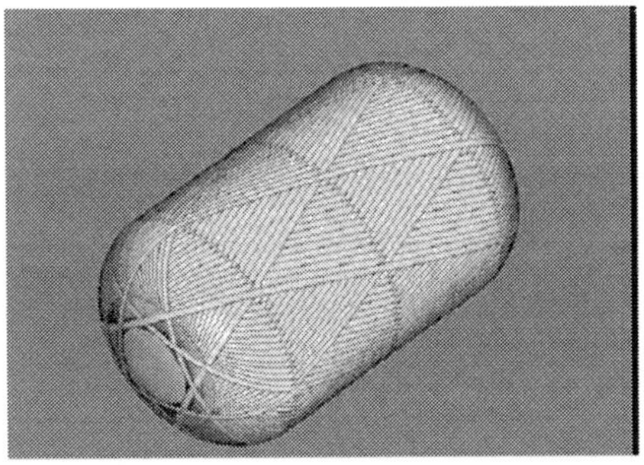

Wound to port

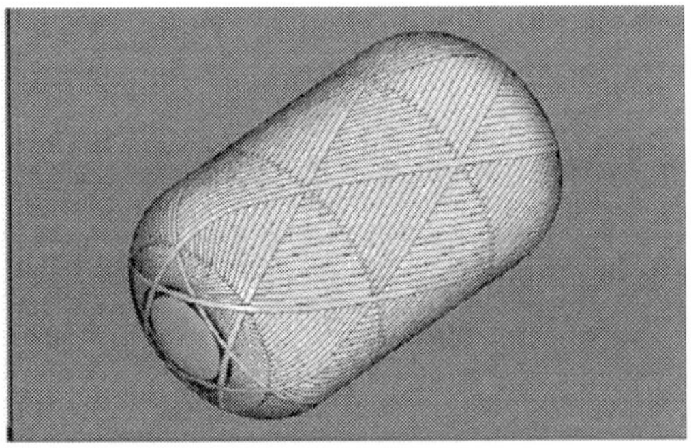

Stepped 1 bandwidth from port

**Fig. 9.18** Typical helical patterns

**Fig. 9.19** Modified geodesic helical winding

**Fig. 9.20** Hoop and helical patterns for constant friction winding

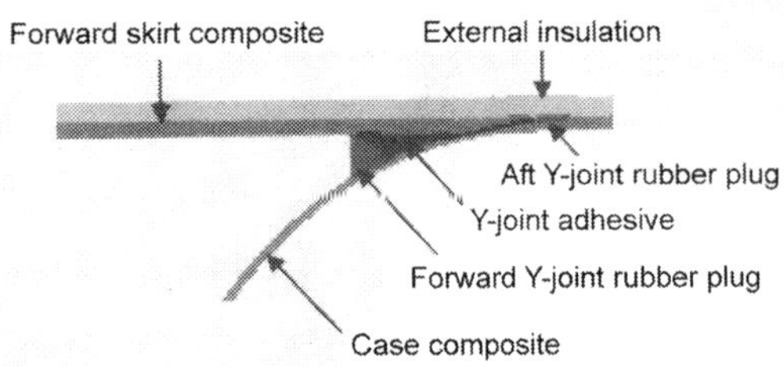


**Fig. 9.21** Skirt joint on rocket motor case

Often, a low-angle cone blends the cylinder with the dome contour in a smooth transition. This provides the interface for the skirt to mate with the case and also minimizes discontinuity stresses. There are several methods of attachment, depending on loads or the preference of the designer/manufacturing teams. Some of these include:

- Precured skirts that are wound cured and machined prior to assembly onto the cured pressure vessel assembly. This approach requires precision bonding tooling to maintain accurate dimensions.
- Wound-in-place skirts, which are co-cured with the pressure vessel. This requires a skirt tool called a skirt dam, which provides support for lay-up and cure and is removed following cure to allow for attachment of metallic flanges.
- Shear plies (Y-joint) are often used for skirt attachment to the pressure vessel. When loads are low, resin bonding can be adequate, although the mismatch in radial and shear deflections causes very high peak shear stresses in the bonds. In skirt attachments with very high acceleration loads, rivets or bolts have also been used.

The composite portion of the skirt consists of hoops, ±45° mats, and longitudinal fibers (unidirectional axial mat or woven cloth and even low-angle helical layers) interspersed to approximate quasi-isotropic properties. The joint between the skirt and the motor case must transfer the combined loading through shear. Often, a shear ply consisting of rubber or film adhesive is laid-up, overwound, and then co-cured with the RMC body, although prefabricated skirts have been bonded to the pressure vessel using adhesive and a rubber or film adhesive shear ply.

### *B-Stage and Cure*

In the first step of the cure cycle, called B-staging, resin viscosity is advanced by means of external heat (lamps or ovens) to the point where cross linking of the epoxide groups is initiated. At this point, the resin is still soft to the touch and still exhibits some tack but will reflow when further heat is applied. The longer the B-stage, the more cross linking of epoxide groups and the less the resin flows, often to the point of becoming tack-free. This allows removal of excess resin prior to proceeding with cure. This is often done with plastic paddles, wiping off the resin runs periodically while the part is rotating until the resin has advanced into the B-staged state. In many RMC applications, the domes are B-staged first, while the cylinder is temporarily covered with insulation to prevent resin advancement. This allows co-curing of the skirt to the cylinder and guarantees a run-free surface for the dome under the skirt tooling. Fabrication of the composite portion of the assembly is completed by the curing operation.

### *Mandrel Removal*

Water-soluble sand mandrels are the easiest to remove. Hot water is added through the wind axis tooling, the sand is washed out, and the wind axis is removed. Removal is more difficult for mandrels where the tooling is swept plaster over a metallic structure or segmented and collapsible. Swept plaster mandrels may require that plaster be chipped out by hand. This operation is laborious and has a high potential for damage to the insulation. When using segmented mandrels, consider approaches to remove each segment without damaging the internal structure of the pressure vessel.

Additionally, if the attach skirts were co-processed with the pressure vessel, the skirt dam tooling must be removed to allow handling and attachment of metallic flanges.

### *Proof Test*

Proof testing has always been part of the acceptance criteria and, in all probability, will continue to be the preferred nondestructive inspection method. Normally, motor cases are proof tested at 0 to 25% higher pressure than their maximum expected operating pressure (MEOP). Minimum burst pressures are normally designed to be 1.25 to 1.75 times MEOP.

Safety factors for CNG, natural gas vehicles, and hydrogen tanks can range as high as 2.25 for carbon-fiber pressure vessels, 3.0 for aramid-fiber pressure vessels, and 3.5 for glass-fiber pressure vessels to allow for pressure cycling.

There is controversy that the proof test actually degrades the case. There is no question that the test produces crazes and cracks within the resin matrix; however, it is questionable if any significant fiber damage results, except for aramid-fiber designs due to the nature of the fiber. Many design engineers believe that some benefits actually result by freeing the cure-induced residual stresses through proof testing. The bottom line is that everyone always feels more comfortable with an RMC that has already seen ≥MEOP to 25% higher pressure than MEOP.

On the other hand, some pressure vessels are not proof tested. The qualification plan relies more heavily on qualification testing, which calls for high numbers of cycle tests to evaluate the fatigue life of the case and liner. A limited number of vessels are cycled, sometimes after conditioning at elevated temperature/humidity, then burst tested and compared against virgin strength. Successful testing can qualify a whole production lot of pressure vessels in lieu of periodic lot sampling testing throughout production.

### *X-Ray Examination*

Most acceptance specifications require x-ray examination. Areas of specific interest are the polar bosses and the skirt junctures. The polar bosses are examined for damage during processing such as change of angle of the boss flange, damage to the shear lip, or damage to threaded connections. The skirt junctures are examined for delaminations or unbonding in the shear ply Y-joint. A decision on when to perform the x-ray is often an important discussion. Most often, the x-ray is completed after the proof test but before filling the internal volume with the required gas, fuel, or propellant.

### *Final Inspection*

The final operation involves physical inspection of the RMC and verification of all of the paperwork that accompanied the part throughout its history. Inspections include critique dimensions, weight, and a visual inspection. The paperwork that accompanied the part has been suggested to weigh more than the part itself. This, of course, is an exaggeration, but the point is that certification includes the entire history of all components, that is, forgings, resin components, fiber lots, and rubber lots. The manufacturing traveler is examined for conformance to cure cycles, processing time, and completeness of procession operations.

## Analysis

The analysis tools commonly used are various finite-element computer codes. This allows users to subject any design that can be configured to both static and dynamic load analyses. Loads include internal pressure, external loads, transient and steady flight loads, and transportation and handling loads, including random vibration.

Critical areas of concern for filament-wound pressure vessels are shown in Fig. 9.22. This figure highlights the areas that cannot be readily analyzed with standard closed-form solutions or relative standard methods such as classical laminate analysis.

Common two-dimensional, axisymmetric finite-element codes, including ABAQUS, NASTRAN, ANSYS, ARGUS, SAASMC, and TEXLSP, can perform accurate evaluations using orthotropic elastic/plastic material properties and large displacement/strain Lagrangian formulation with special "gap" elements for interfaces, automatic mesh generation, and extensive postprocessing, including graphic display.

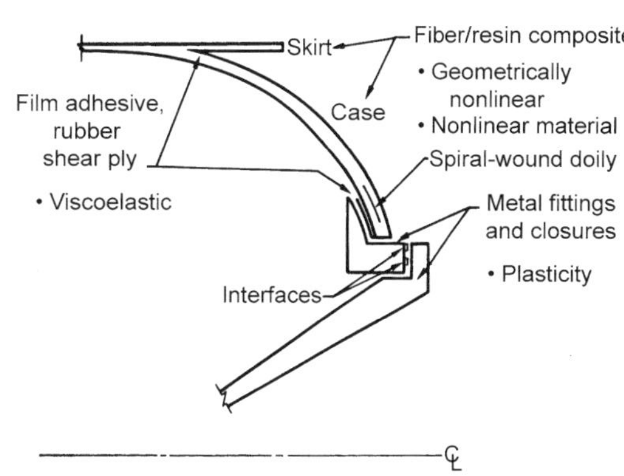


**Fig. 9.22** Critical regions for sophisticated analysis techniques

Common three-dimensional, dynamic finite-element codes include SPAR, NASTRAN, ABAQUS, and ANSYS. These codes can handle the orthotropic elastic properties and perform natural frequency/mode shape response and predict response to a variety of dynamic loads.

Figure 9.23 shows a typical finite-element grid for the analysis of a typical pressure vessel. Figure 9.24 shows a typical response to internal pressurization for a typical pressure vessel design.

A typical comparison between predicted and measured strain response is shown in Fig. 9.25. These comparisons are for an intermediate-modulus carbon-fiber/epoxy pressure vessel.

A typical comparison between predicted and measured strain response is shown in Fig. 9.26.

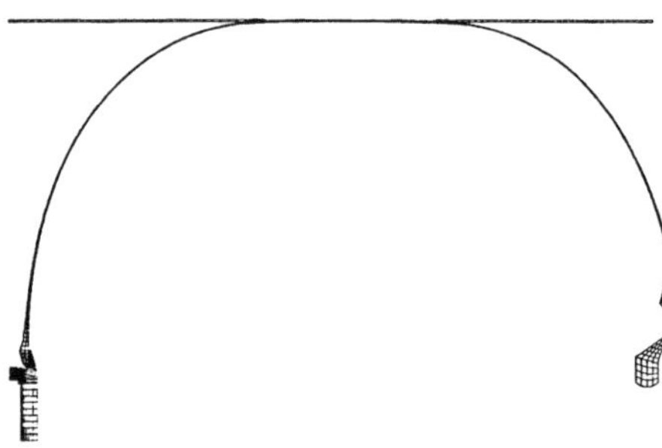

**Fig. 9.23** Finite-element model of a typical pressure vessel

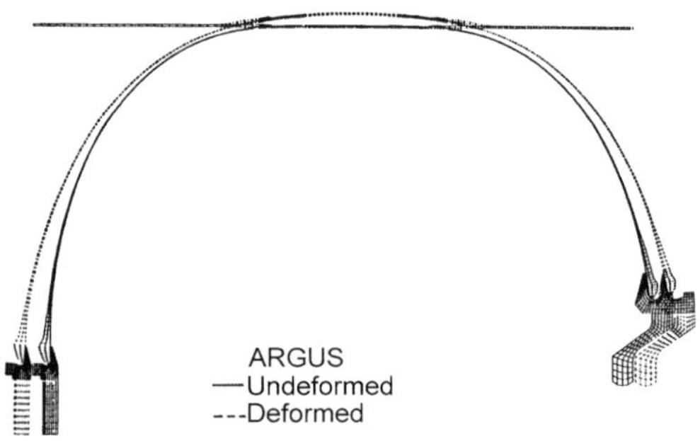


**Fig. 9.24** Pressurized deflections for a typical pressure vessel

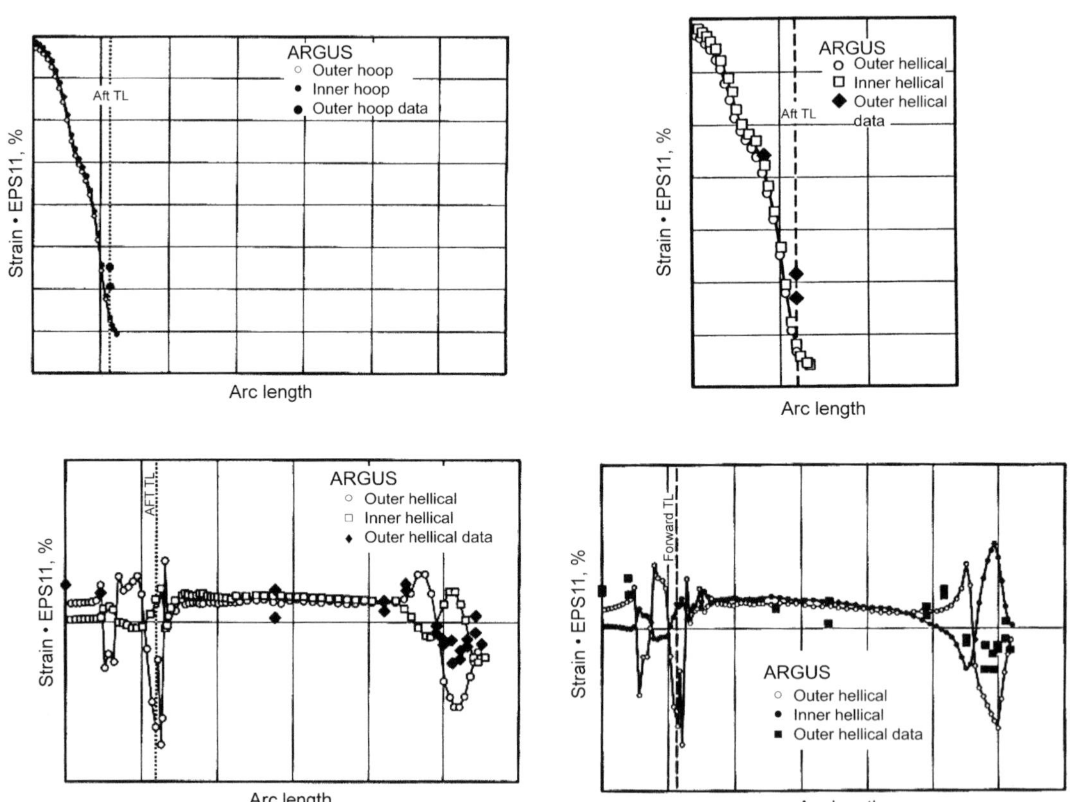


**Fig. 9.25** Measured vs. predicted strain comparisons for carbon-fiber pressure vessel

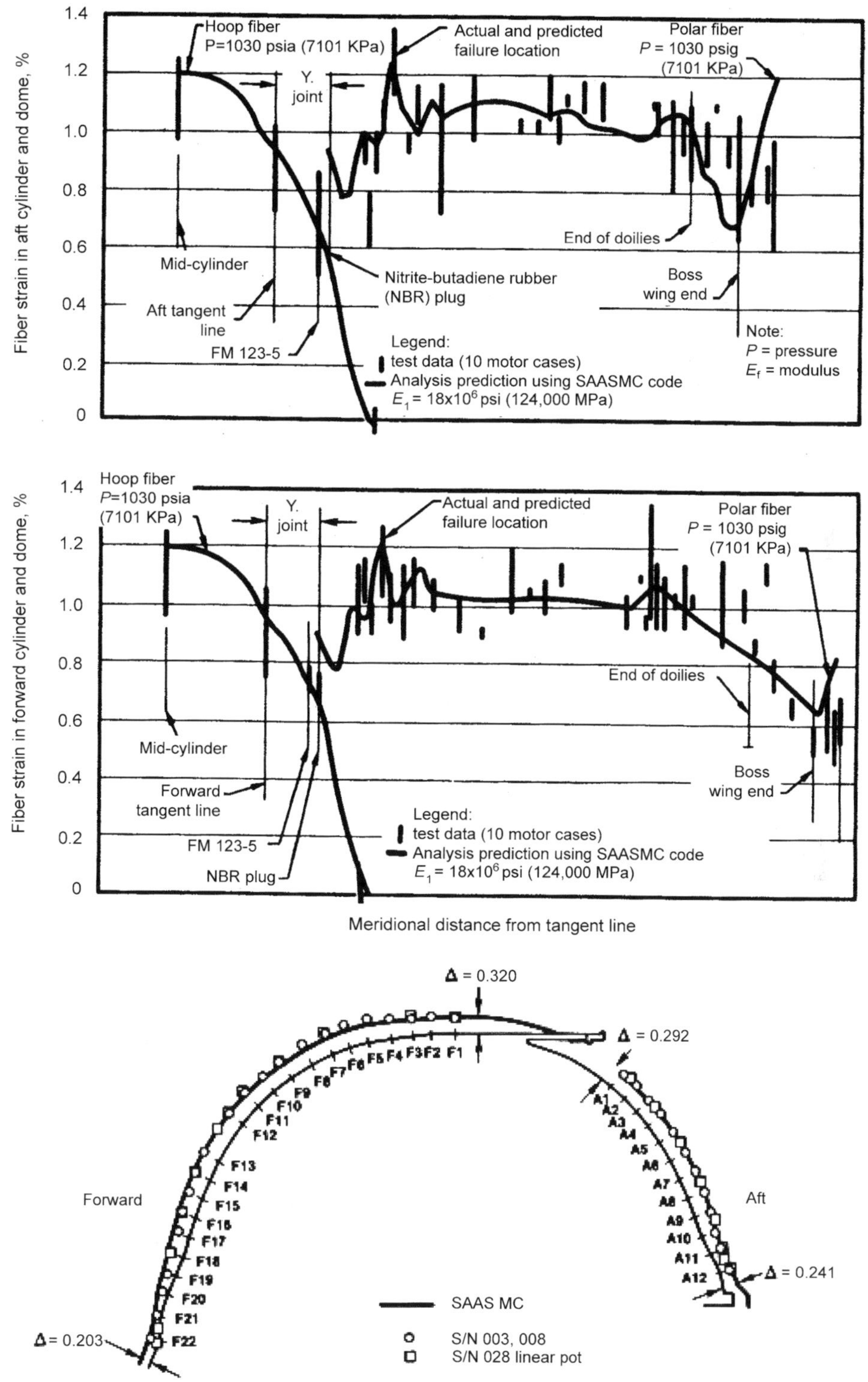


**Fig. 9.26** Predicted vs. measured response for Kevlar 49 fiber pressure vessel

These comparisons are for an aramid-fiber/epoxy pressure vessel.

## Testing

Testing is critical to successful composite pressure vessel design. The designer must know how the various materials can be expected to perform, and this usually means testing. Testing encompasses the raw materials as well as subscales, analogs, and the full-scale hardware. Table 9.8 shows the typical testing required for pressure vessels.

The industry tries to standardize testing as much as possible so that there is a common frame of reference for comparing competitive designs. ASTM International and Joint Army, Navy, NASA, Air Force (JANNAF) publish two of the most common testing standards. Rarely are these standards used without some form of modification required due to facility, tooling, or available hardware differences. Analysis of the test specimens using the finite-element modeling technique is often used to generate the properties necessary for the final product design.

### *Materials*

Raw materials often need characterization, especially when they are newly developed. In recent years, advanced fibers have become available, and testing has been focused on characterizing the new fibers with existing well-understood resin systems before trying to develop new resin systems. Resin system development has been driven in recent years by environmental and safety concerns more than performance needs. Understanding how the materials react to the manufacturing process, either at the pressure vessel level or the finished assembly level, is mandatory because the behavior of the design must be qualified for the actual operational environment.

**Table 9.8 Testing for pressure vessel composites**

- Standards common to composite testing
  - ASTM International
  - Suppliers of Advanced Composite Materials Association (SACMA)
  - Joint Army, Navy, NASA, Air Force (JANNAF)
- Materials
  - Fundamental properties of fibers and resins
  - Simple composite laminates
  - Tension, compression, flexure, shear vs. temperature, including moisture and aging effects
  - Strength of materials, close-form calculations
- Subscale and analogs
  - Bottles, rings, short beams, cylinders with and without joints
  - Tension, compression, shear vs. temperature, including moisture and aging effects
  - Derive allowable strengths using analytical models and strength of materials, close-form calculations
- Full scale
  - Development testing for burst and structural loads to validate structural models and manufacturing process
  - Production reproducibility testing (also called performance evaluation testing)
  - Motor testing
  - Insensitive munitions testing (fast and slow cookoff, bullet impact, shaped-charge impact, etc.)
  - Transportation and handling, including drop testing

### *Composite Characterization*

Composite testing (characterization and aging testing) consists of laminates and analog test specimens that mimic the application and design characteristics. These include small pressure vessels (5.75 in., or 146 mm, in diameter) in accordance with ASTM D 2585 or modified versions such as Pratt & Whitney Space Propulsion's 12 in. (305 mm) diameter pressure vessel. These are used for delivered fiber strength versus helical/hoop stress ratio, including the effects of local or global reinforcements as well as the effects of motor processing, aging, and temperature. Other companies have used sizes up to 37 in. (940 mm) for generic bottles. Fractional-scale small pressure vessel analogs or full-scale rings are also often used. In addition, Naval Ordnance Laboratory ring short-beam interlaminar shear and Iosipescu transverse shear testing versus temperature and moisture are frequently used to characterize the fiber and resin. The following are used to demonstrate specific design capabilities:

- Cylinder-oriented, 4.0 in. (102 mm) JANNAF tube analog for compression testing versus temperature, moisture, and aging
- Skirt-oriented, 4.0 in. (102 mm) JANNAF tube analog for compression testing versus temperature, moisture, and aging
- Skirt-to-case Y-joint, 4.0 in. (102 mm) JANNAF tube analog for shear and compression testing
- Flange-to-skirt riveted joint, 4.0 in. (102 mm) JANNAF tube analog for joint tension or compression testing versus temperature, moisture, and aging
- Dome-oriented, 12.0 in. (305 mm) flat panel wound for $\pm\alpha$ panels for tensile and compression testing, often to support manufac-

turing issues, curved panels, and large cylinders used for insulation or liner-to-case bond evaluation (bond-in-tension, lap shear, 90° or 180° peel) versus temperature, moisture, and aging, including motor process effects

Typical data comparisons can be made for predicted and measured properties. The agreement is pretty good most of the time (Table 9.9). The best agreements are for fiber-dominated properties. Representative examples are for a typical intermediate-modulus carbon-fiber/epoxy resin and for aramid-fiber/epoxy resin laminates.

Composite materials characterization for insulated pressure vessels involves the use of analog and subscale test specimens that duplicate the structural element that is being evaluated, as shown in Fig. 9.27. Full-thickness,

**Table 9.9 Predicted versus measured properties for carbon- and aramid-fiber composites**

| Parameter | Predicted T-800/epoxy | Measured T-800/epoxy | Predicted Kevlar 49/epoxy | Measured Kevlar 49/epoxy |
|---|---|---|---|---|
| $E_{11}$, $10^6$ psi (GPa) | 25.414 (175.227) | 25.644 (177.195) | 12.4 (85.495) | 12.1 (83.427) |
| $E_{22}$, ksi (GPa) | 1294.973 (8.963) | 1299.2–1499.3 (8.96–10.34) | 749.845 (5.17) | 749.845 (5.17) |
| $\nu_{12}$ | 0.26 | 0.33 | 0.36 | 0.34 |
| $G_{12}$ ksi (GPa) | 490 (3.378) secant | 710 (4.895) initial | 270 (1.862) | 300 (2.068) |
| $FTU_{11}$, ksi (MPa) | 480 (3309.5) | 488 (3364.6) | 200 (1378.9) | 200 (1378.9) |
| $FCU_{11}$, ksi (MPa) | 142 (–979.0) | 250 (–1723.7) | 40 (–275.8) | 32 (–220.6) |
| $FTU_{22}$, ksi (MPa) | 1900 (13.1) | 2891 (19.93) | 4293 (29.6) @ 60% fiber volume | 615–1798 (4.24–12.4) |
| $FSU_{ILS}$, ksi (MPa) | 6498–9993 (44.8–68.9) | 9297–13996 (64.1–96.5) | 6295 (43.4) @ 60% fiber volume | 3495–5004 (4.24–12.4) |
| CTE, mm/mm/°C | | | | |
| 11 | $-1.21 \times 10^{-7}$ | $-7.2 \times 10^{-8}$ | $-3.96 \times 10^{-6}$ | $-2.88 \times 10^{-6}$ |
| 12 | $36 \times 10^{-6}$ | $27 \times 10^{-6}$ | $57.6 \times 10^{-6}$ | $49.5 \times 10^{-6}$ |
| Volume fraction | 0.60 | 0.60–0.62 | 0.65 | 0.65–0.70 |

ASTM D 2585 modified (increased diameter) 305 mm pressure vessel

102 mm JANNAF tubes

ASTM short beam shear (cut from pressure vessel)

102 mm JANNAF compression test tubes

**Fig. 9.27** Typical analog test specimens for characterization and aging. JANNAF: Joint Army, Navy, NASA, Air Force

small-diameter test specimens can be analyzed using the same finite-element programs that are used for the full-scale case and therefore develop strength and stiffness allowables that can be used at the full-scale level using the validated analysis models.

### Full-Scale Testing

Full-scale testing is often conducted to confirm compliance to design requirements and to calibrate analytical models. When conducting full-scale testing, instrumentation is very important. Instrumentation for structural testing is expensive and time-consuming. It is, however, the only way to validate the analysis models and be able to look at load conditions not tested. A typical instrumentation list is shown in Table 9.10. Burst test examples are shown in Fig. 9.28 for a variety of design configurations.

## Expected Performance

Performance parameters for composite motor cases include fiber strength and how much it weighs for the loads it can take. This is called pressure vessel efficiency, or *PV/W.*

The delivered strength was presented earlier in this chapter in Fig. 9.10. Total composite case efficiency *PV/W* (including the skirts, Y-joints, polar boss, and flanges) is somewhat geometry-driven, with the best values demonstrated on high length-to-diameter (*L/D*) ratio composite case designs, as expected. Figure 9.29 shows the total case *PV/W.* This is a good estimation tool for new motor designs. Full open ends penalize the total case *PV/W*, as shown in Fig. 9.29 for the IM-7R/T-800 case at an *L/D* of 10, because of the large diameter, increased composite usage, and heavy metallic structures needed for this type of design (Ref 9.12, 9.14–9.16).

**Table 9.10 Typical instrumentation for full-scale structural testing**

- **Pressure**
  Pressure gage: 2
  Pressurizing rate: >100 psi
- **Loads**
  Up to five external load jacks
  Load rate at 1% of limit
- **Strain**
  Dome helical fiber direction (120° apart): 77 (depending on diameter)
  Cylinder hoop fiber direction (120° apart): 21 (depending on length)
  Cylinder axial or transverse direction or rosettes (120° apart): 21 (depending on length)
  End fitting hoop (90° apart): 4
- **Displacement gages (typically removed for burst)**
  Aft and forward end fitting axial: 3 each (120° apart), used for axial displacement and tilt
  Aft end fitting radial: 2 (90° apart), used for thrust offset
  Cylinder hoop wire: Up to 3 (depending on length)
  Dome/cylinder normal/radial: 136 (depending on diameter)
- **Acoustic:** 2 microphones in cylinder 180° apart (depending on length)
- **Video:** Up to 4 cameras with frame rates from 64–1000 fps (depending on length)
- **High-speed camera (1000/3000 fps):** Up to 10 cameras (depending on size)

### REFERENCES


9.1 B. Wilson, *SAMPE J.*, May/June 1997, p 25–32

9.2 G.P. Sutton and D.M. Ross, *Rocket Propulsion Elements*, John Wiley & Sons, 1976

9.3 D. Gutierrez-Lemini, J. Hurdle, C. Kawabata, H. Choi, C. Lemler, and P. Empleo, "Composite Design and Analysis Code," developed for the U.S Air Force Rocket Propulsion Laboratory by United Technologies Corporation Chemical Systems Division, San Jose, CA, 1985

9.4 A. Wozchiechowicz, "Design of Filament Wound Rocket Cases," ABL Technical Report, Dec 1968 (netting analysis and dome contouring)

9.5 S.T. Peters, W.D. Humphrey, and R.F. Foral, *Filament Winding Composite Structure Fabrication*, 2nd ed., SAMPE Publishers, Covina, CA, 1998

9.6 "Test Method for Preparation and Tension Testing of Filament-Wound Pressure Vessels," D 2585-68, ASTM, 1990 (withdrawn in 1996 but still used by most pressure vessel companies)

9.7 K.C. Gramoll, F. Namiki, and J. Onoda, Dome Thickness of Filament Wound Pressure Vessels, *Trans. Jpn. Soc. Aeronaut. Space Sci.*, Vol 33 (No. 100), 1990, p 66–79

9.8 R.W. Yeager and J.R. Hinchman, *20th Technical Conf. Proc.*, Society of Plastic Engineers, 1964

9.9 S. Timoshenko and S. Woinowsky-Krieger, *Theory of Plates and Shells*, 2nd ed., McGraw-Hill Book Co., 1959 (membrane theory)

9.10 J.-P. Denost, "New Design Concepts for Filament-Wound Pressure Vessel with

**Fig. 9.28** Pressure vessel burst test examples. (a) 92 in. (2336 mm) large aramid-fiber/epoxy space motor case burst. (b) S-glass pressure vessel burst. (c) Tactical rocket motor carbon-fiber/epoxy burst from high-speed film. (d) Carbon-fiber/epoxy space motor pressure vessel burst on high-speed film

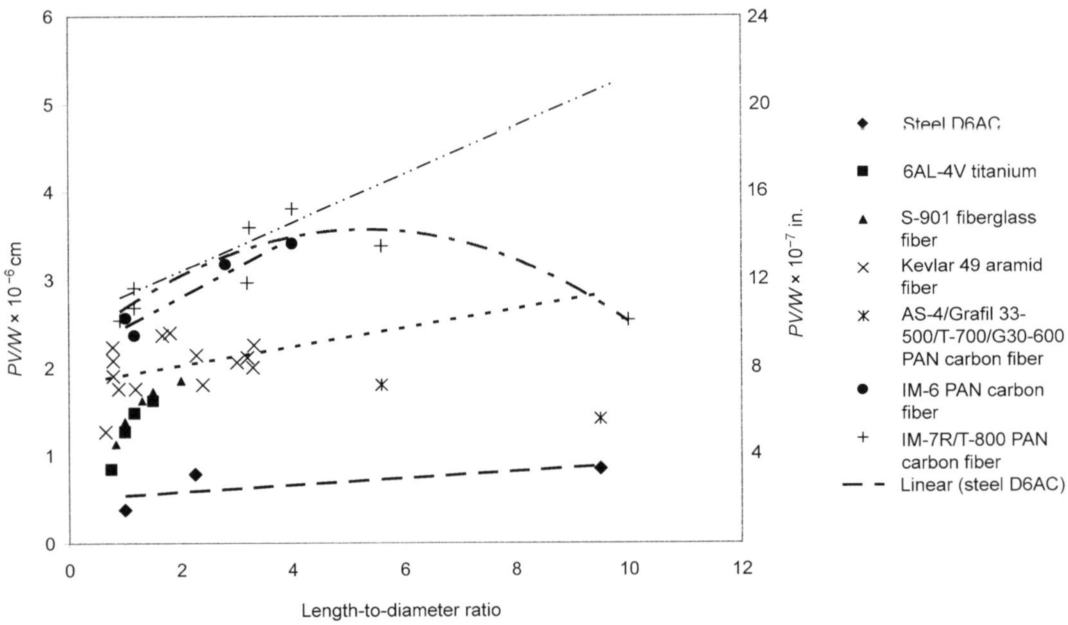


**Fig. 9.29** Total pressure vessel efficiency, *PV/W*, for representative materials. *P*, pressure; *V*, volume; *W*, weight; PAN, polyacrylonitrile


Unequal Polar Openings," AIAA-1982-1067, AIAA/SAE/ASME 18th Joint Propulsion Conference, June 21–23, 1982 (Cleveland, OH)

9.11 J.C. Halpin, *Primer on Composite Materials: Analysis*, Technomic Publishing Company, revised 1984

9.12 H.M. Reynolds and M. Dhillon, "Development of the Orbus-1 Carbon Fiber/Epoxy Motor Case for the STARS/STARBIRD Programs," presented at the 1990 JANNAF Propulsion Conference (Monterey, CA)

9.13 S. Timoshenko, *Strength of Materials, Part II: Advanced Theory and Problems*, 3rd ed., Robert F. Krieger Publishing Co., Huntington, NY

9.14 H.M. Reynolds, "Composite Case Design and Analysis, The-State-of-the-Art," presented at the 39th AIAA Joint Propulsion Conference, July 20–23, 2003 (Huntsville, AL)

9.15 H.M. Reynolds, "Composite Case Design and Modeling," presented at the 52nd JANNAF Propulsion Conference, May 10–13, 2004 (Las Vegas, NV)

9.16 H.M. Reynolds, "Motor Case Design," presented for the Advanced Solid Rocket Propulsion Course MAE 695, University of Alabama, Huntsville, Feb 2007

Copyright © 2011, ASM International®
All rights reserved.
www.asminternational.org

# APPENDIX 1

# Specifications and Designations of Fibers and Resins

## Fibers

| Fiber designation | Type | Commercial source |
|---|---|---|
| AS-4 | High-strength carbon/graphite | Hexcel |
| T-400 | Carbon/graphite | Toray, SGL |
| T-650 | Carbon/graphite | Toray |
| IM-7 | Intermediate-modulus carbon/graphite | Toray |
| M46J | High-modulus carbon/graphite | Toray |
| OC 463 750 yield | E-glass fiber tow | Owens Corning |
| M40J | Carbon/graphite | Toray |
| T 1000G | Highest-strength carbon/graphite | Toray |
| T-300 | High-strength carbon/graphite | Toray |
| T-700 | Carbon/graphite | Toray |
| T-800 | Carbon/graphite | Toray |
| S2 glass | High-strength glass tow | AGY |
| Zylon HM | Poly(p-phenylene-2,6-benzobisoxazole), high-performance fibers | Toyobo |

## Resins

| Resin designation | Type | Supplier |
|---|---|---|
| 3501 | Epoxy prepreg, 350 °F cure | Hexcel Aerospace |
| LY564 | Epoxy | Araldit/Huntsman |
| 1902 | Epoxy prepreg, 250 °F cure | ... |
| 8552 | Epoxy prepreg, 350 °F cure | Hexcel Aerospace |
| Udel P1700 | Thermoplastic resin | Solvay |
| RF007 | Resin | ... |
| Epon 826 | Liquid epoxy, bisphenol-A, low viscosity | Hexion |
| HBRF 55 | Liquid epoxy, amine-cured system | Proprietary formulation |
| Epon 9405 | Liquid epoxy resin | Unknown |

Copyright © 2011, ASM International®
All rights reserved.
www.asminternational.org

# APPENDIX 2

# Glossary of Filament-Winding Terms

## A

**adhesive.** A substance capable of holding two materials together by surface attachment. Adhesive can be film, liquid, or paste.

**aging.** The effect of materials of exposure to an environment for an interval of time. The process of exposing materials to an environment for an interval of time.

**amine resin.** A synthetic resin derived from the reaction of urea, thiourea, melamine, or allied compounds with aldehydes, particularly formaldehyde.

**angle of winding (or wind angle).** The angle at which the roving band is laid with respect to the mandrel axis of rotation, circumferential winding being approximately 90°.

**anisotropic.** Not isotropic; exhibiting different properties when tested along axes in different directions.

**aramid.** Generic name for poly para-phenylene terphlalamide organic fibers, which include Kevlar and Nomex. See also *Kevlar.*

**A-stage.** An early stage in the polymerization reaction of certain thermosetting resins (especially phenolic) in which the material, after application to the reinforcement, is still soluble in certain liquids and is fusible. Also called resole. See also *B-stage* and *C-stage.*

**axial winding.** In filament-wound reinforced plastics, a winding with the filaments parallel to, or at a small angle to, the axis (0° helix angle).

## B

**balanced design.** In filament-wound reinforced plastics, a winding pattern so designed that the stresses in all filaments are equal.

**balanced-in-plane contour.** In a filament-wound part, a head contour in which the filaments are oriented within a plane and the radii of curvature are adjusted to balance the stresses along the filaments with the pressure loading.

**band density.** In filament winding, the quantity of fiberglass reinforcement per inch of band width, expressed as strands (or filaments) per inch.

**band width.** In filament winding, the width of the reinforcement as it is applied to the mandrel (measured perpendicular to the band).

**bismaleimide (BMI).** A type of polyimide that cures by an addition rather than a condensation reaction (thus avoiding problems with volatiles formation), which is produced by a vinyl-type polymerization of a prepolymer terminated with two maleimide groups. They are intermediate in temperature capability between epoxy and polyimides.

**bleedout.** In filament winding, the excess liquid resin that migrates to the surface of a winding.

**boss.** Commonly referred to as polar boss or end fitting. It is a metal fitting located in the center of each dome that describes the pole about which winding bands are wrapped.

**bottle winds.** These combine a helical wind section with a geodesic, or near-geodesic, wind pattern to generate a stable fiber path over the end domes of the part. These are typically used for pressure vessels.

**B-stage.** An intermediate stage in the reaction of certain thermosetting resins in which the material swells when in contact with certain liquids and softens when heated, but may not entirely dissolve or fuse; sometimes referred to as resitol. The resin in an uncured prepreg or premix is usually in this stage.

**bulk factor.** Reciprocal of fiber volume. Used to calculate lamina thickness.

**burst strength.** Hydraulic pressure required to burst a vessel of given thickness. Commonly used in testing filament-wound composite structures.

## C

**carbon/graphite fiber.** Fiber produced by the pyrolysis of organic precursor fibers, such as rayon, polyacrylonitrile (PAN), and pitch, in an inert environment. The term is often used interchangeably with the term *graphite*; however, carbon fibers and graphite fibers differ. The basic differences lie in the temperature at which the fibers are made and heat treated, and in the amount of elemental carbon produced. Carbon fibers typically are carbonized in the region of 2400 °F (1315 °C) and assay at 93 to 95% C, while graphite fibers are graphitized at 3450 to 4500 °F (1900 to 2480 °C) and assay at more than 99% elemental carbon.

**catalyst.** A substance that changes the rate of a chemical reaction without itself undergoing a permanent change in composition or becoming a part of the molecular structure of the product. A substance that markedly speeds up the cure of a compound when a minor quantity is added, compared to the amounts of primary reactants.

**catenary.** The tendency of some strands in a taut horizontal roving to sag lower than the others.

**chain/transition winds.** The user can combine various individual winds to form multiple layers of a complex wind. The software will also attempt to generate stable transitions from one layer to the next.

**circuit.** One complete traverse back and forth of the fiber-feed mechanism of a winding machine.

**circumferential ("circ") winding.** In filament-wound reinforced plastics, a winding with the filaments essentially perpendicular to the horizontal mandrel axis (90° or hoop winding).

**coefficient of variation.** The ratio of the standard deviation to the mean.

**compaction.** In reinforced plastics and composites, the application of a temporary press bump cycle, or vacuum, to remove trapped air and compact the lay-up. Pressure exerted on previously wound, uncured layers by high-angle or circumferential windings.

**composite material.** A combination of two or more materials (reinforcing elements, fillers, and composite matrix binder) differing in form or composition on a macroscale. The constituents retain their identities: They do not dissolve or merge completely into one another although they act in concert. Normally, the components can be physically identified and exhibit an interface between one another.

**continuous filament.** An individual rod of small-diameter fiber that is flexible and of great or indefinite length.

**continuous filament yarn.** Yarn formed by twisting two or more continuous filaments into a single, continuous strand.

**creel.** A device for holding the required number of roving balls or supply packages in a desired position for unwinding onto the mandrel when filament winding.

**C-stage.** The final stage in the reaction of certain thermosetting resins in which the material is practically insoluble and infusible. Sometimes referred to as resite. The resin in a fully cured thermoset molding is in this stage. See also *A-stage* and *B-stage*.

**cure.** To irreversibly change the properties of a thermosetting resin by chemical reaction (that is, condensation, ring closure, or addition). Cure may be accomplished by addition of curing (cross-linking) agents, with or without heat and pressure.

**cure cycle.** The time/temperature/pressure cycle used to cure a thermosetting resin system.

## D

**delamination.** The separation of a laminate along the plane of its layers. Also the separation of bonded insulation within the adhesive layer or at the adhesive interface.

**denier.** A yarn and filament numbering system in which the yarn number is equal to the weight in grams of 9000 m. Used for continuous filaments. The lower the denier, the finer the yarn.

**diameters.** *Outside diameter:* The outside major diameter of a cylindrical part. *Inside diameter:* The inside major diameter of a cylindrical part. This would correspond to the outside diameter of the mandrel. *Mean diameter:* The diameter halfway between the outside and inside diameters. *Major diameter:* The largest diameter of a tapered or nonuniform mandrel or part. *Minor diameter:* The smallest diameter of a tapered or nonuniform mandrel or part. *Pole* *diameter:* The minimum diameter at the end of a closed-end vessel. *Sprocket diameter:* The pitch diameter of the carriage drive sprocket. *Eye diameter:* The inside diameter of round or semicircular deliver eyes. The inside surface is usually toroidal in shape.

**doctor blade (or bar).** A straight piece of material (usually smooth metal) used to spread and control the amount of resin applied to the roving, tow tape, or fabric.

**doily.** A planar reinforcement applied to local areas (usually around port openings) to provide additional strength.

**dome.** In a cylindrical container, the portion that forms the integral ends of the container.

**doubler.** A local area with extra wound reinforcement, wound integrally with the part or wound separately and fastened to the part.

**dry winding.** A process in which preimpregnated B-staged fibers are used in the winding.

## E

**E-glass.** A borosilicate glass, the type most used for glass fibers for reinforced plastics.

**end.** A strand of roving consisting of a given number of filaments gathered together (the group of filaments is considered an end or strand before twisting and a yarn after twist has been applied); an individual warp yard, thread, fiber, or roving.

**end count.** An exact number of ends supplied on a ball or roving.

**epoxide equivalent weight (EEW).** The weight of a resin in grams that contains one gram equivalent of epoxy.

**equator.** In a pressure vessel, the juncture plane between the cylindrical portion and the end dome. Also called the tangent line.

**even tension.** Describes the process whereby each end of roving is kept in the same degree of tension as the other ends making up that ball of roving. See also *catenary.*

**extractor.** A machine used to pull off the mandrel through a small hole, stripping the part that was wet-wound onto the mandrel. A method of part removal.

**eye height.** The distance between the delivery eye and the mandrel surface (measured in a radial direction from the mandrel axis).

## F

**fatigue.** The failure or decay of mechanical properties after repeated applications of stress. Fatigue tests give information on the ability of a material to resist the loss of strength from repetitive loading.

**fiber.** A general term used to refer to filamentary materials. Often, *fiber* is used interchangeably with *filament.* It is a general term for a filament with a finite length that is at least 100 times its diameter, which is typically 0.004 to 0.005 in. (0.10 to 0.13 mm). (In most cases, it is prepared by drawing from a molten bath, spinning, or deposition on a substrate. Fibers can be continuous or specific short lengths, that is, discontinuous, normally no less than ⅛ in., or 3.2 mm). See the Appendix for specifications and designations of fibers.

**fiberglass.** An individual filament made by attenuating molten glass. A continuous filament in a glass fiber of great or indefinite length.

**filaments.** Individual fibers of indefinite length.

**filament winding.** A process for fabricating a composite structure in which continuous reinforcements (filament, wire, yarn, tape, or other), previously impregnated with a matrix material or impregnated during the winding, are placed over a rotating and removable form or mandrel in a previously prescribed way to meet certain stress conditions. Generally, the shape is a surface of revolution and may or may not include end closures. When the right number of layers is applied, the wound form is cured and the mandrel removed.

**finish.** A material applied to the surface of fibers in a roving or fabric used to reinforce plastics and intended to improve the physical properties of such reinforced plastics over that obtained using reinforcement without finish.

**flow.** The movement of resin under pressure, allowing it to fill all parts of a mold; the gradual but continuous distortion of a material under continued load, usually at high temperature. Also known as creep.

**fracture toughness.** A measure of the damage tolerance of a material containing initial flaws or cracks. G1c and G2c are the critical strain-energy release rates in the 1 and 2 directions.

**fuzz.** Accumulation of loose, broken filaments (usually the result of abrasion).

## G

**G1c, G2c.** See *fracture toughness.*

**gap (filament winding).** The space between successive windings, which are usually intended to lay next to each other.

**gel.** A semisolid system consisting of a network of solid aggregates in which liquid is held; the initial gellike solid phase that develops during the formation of a resin from a liquid.

**gel coat.** A quick-setting resin applied to the surface of a mold and gelled before lay-up. The gel coat becomes an integral part of the finished laminate and is usually used to improve surface appearance and bonding.

**geodesic.** The shortest distance between two points on a surface.

**geodesic-isotensoid contour.** In a filament-wound reinforced plastic pressure vessel, a dome contour in which the filaments are placed on geodesic paths so that the filaments, when wet, will not slip on the mandrel surface and will exhibit uniform tension throughout their length under pressure loading.

**glass.** An inorganic product of fusion that has cooled to a rigid condition without crystallizing. (Glass is typically hard and relatively brittle and has a conchoidal fracture.)

**graphite fiber.** A fiber made from a precursor by an oxidation, carbonization, and graphitization process (which provides a graphitic structure). See also *carbon/graphite fiber.*

## H

**hardener.** A substance or mixture added to a plastic composition to promote or control the curing action by taking part in it.

**helical winding.** In filament-wound items, a winding in which a filament band advances along a helical path, not necessarily at a constant angle, except in the case of a cylinder.

**hoop fibers.** Parallel fibers at nearly 90° to the component centerline.

**hoop stress.** The circumferential stress in a material of cylindrical form subjected to internal or external pressure.

**hybrid.** A composite laminate consisting of laminae of two or more composite material systems. A combination of two or more different fibers, such as carbon and glass or carbon and aramid, in a structure. Tapes, fabrics, and other forms may be combined; usually only the fibers differ. See also *interply hybrid* and *intraply hybrid.*

## I

**impregnate.** In reinforced plastics, the saturation of the reinforcement with a resin.

**interlaminar shear.** Shearing force tending to produce a relative displacement between two laminae in a laminate along the plane of their interface.

**interpenetrating network (IPN).** A combination of two polymers in a network in which at least one (or both) is synthesized and/or cross linked in the presence of the other.

**interply hybrid.** A reinforced plastic laminate in which adjacent laminae are composed of different materials.

**intraply hybrid.** A reinforced plastic laminate in which different materials are used within a specific layer or band.

**isotropic.** Having uniform properties in all directions. The measured properties of an isotropic material are independent of the axis of testing.

## J

**joint.** The location at which two adherends are held together. The general area of contact for a bonded, bolted, or riveted structure.

**joint lap.** A joint made by cutting away similar angular segments of two adherends and bonding them with the cut areas fitted together.

## K

**Kevlar.** An organic polymer, in fiber form, composed of aromatic polyamides having a para-type orientation (parallel chain with bonds extending from each aromatic nucleus).

**knuckle area.** The area of transition between sections of different geometry in a filament-wound part; for example, where the skirt joins the cylinder of the pressure vessel. Also called Y-joint.

## L

**lamina.** A single ply or layer in a laminate. A flat or curved surface containing fabric or unidirectional fibers.

**laminate (noun).** A product made by lamination.

**laminate (verb).** To unite sheets of material by a bonding material, usually with pressure and heat.

**lap.** In filament winding, the amount of overlay between successive windings, usually intended to minimize gapping.

**lap joint.** A joint made by placing one adherend partly over another and bonding the overlapped portions.

**L/D ratio.** A term to define a pressure vessel; denotes the ratio of chamber length to chamber diameter.

**liner.** In a filament-wound pressure vessel, the continuous, usually flexible coating on the inside surface of the vessel, used to protect the laminate from chemical attack or to prevent leakage under strain. Pressure vessel liners are often metal, rubber, or thermoplastic.

**longos.** Low-angle helical or longitudinal windings. Also called longitudinals.

## M

**mandrel.** The form (usually cylindrical) around which resin-impregnated fiber is wound to form pipes, tubes, or vessels.

**mat.** A sheet formed by filament winding a single-hoop ply of fiber on a mandrel, cutting across its width and laying out a flat sheet.

**matrix.** See *resin.*

**mechanical properties.** The properties of a material, such as modulus and compressive and tensile strengths, that are associated with elastic and inelastic reaction when force is applied. The individual relationship between stress and strain.

**moisture absorption.** The pickup of water vapor from air by a material. It relates only to vapor withdrawn from the air by a material and must be distinguished from water absorption, which is the gain in weight due to the takeup of water by immersion.

**mold.** The cavity or matrix into or on which the plastic composition is placed and from which it takes form. To shape plastic parts or finished articles by heat and pressure. The assembly of all the parts that function collectively in the molding process.

**mold-release agent.** A liquid or powder used to prevent sticking to the mandrel.

**monofilament.** A single fiber or filament of indefinite length generally produced by extrusion; a continuous fiber of sufficient size to serve as yarn in normal textile operations. Also called monofil.

**multifilament yarn.** A multitude of fine, continuous filaments (often 5 to 100 individual filaments), usually with some twist in the yarn to facilitate handling. Sizes range from 5 to 10 denier up to a few hundred denier. Individual filaments in a multifilament yarn are usually approximately 1 to 5 denier.

## N

**netting analysis.** The analysis of filament-wound structures that assumes the stresses induced in the structure are carried entirely by the filaments; the strength of the resin is neglected. It also assumes that the filaments possess no bending or shearing stiffness and carry only the axial tensile loads.

**NOL ring.** A parallel filament- or tape-wound hoop test specimen developed by the Naval Ordnance Laboratory (NOL, now the Naval Surface Weapons Laboratory) for measuring various mechanical strength properties of the material, such as tension and compression, by testing the entire ring or segments of it.

**nondestructive inspection (NDI).** A process or procedure, such as ultrasonic or radiographic inspection, for determining the quality or characteristics of a material, part, or assembly without permanently altering the subject or its properties. Used to find internal anomalies in a structure without degrading its properties. Also considered synonymous with nondestructive evaluation (NDE).

## O

**one-circuit winding.** See *single-circuit winding.*

**orthotropic.** Having three mutually perpendicular planes of elastic symmetry.

**overtravel.** The additional carriage or eye travel beyond the ends of the part mandrel that is necessary to provide laydown of the fiber on the mandrel.

## P

**package.** Yarn, roving, and so forth in the form of units capable of being unwound and suitable for handling, storing, shipping, and use.

**peel ply.** A layer of open-weave material, usually fiberglass or heat-set nylon, applied directly to the surface of a prepreg lay-up. The peel ply is removed from the cured laminate immediately before bonding operations, leaving a clean, resin-rich surface that needs no further preparation for bonding, other than application of a primer where one is required.

**permeability.** The passage or diffusion (or rate of passage) of a gas, vapor, liquid, or solid through a barrier without physically or chemically affecting it.

**plastic.** A material that contains as an essential ingredient an organic polymer of large molecular weight, as well as hardeners, fillers, reinforcements, and so forth; is solid in its finished state; and, at some stage in its manufacture or its processing into finished articles, can be shaped by flow. A plastic may be either thermoplastic or thermoset.

**ply.** A single layer of prepreg. A single pass in filament winding (two plies forming one layer).

**PMR polyimides.** A novel class of high-temperature-resistant polymers. PMR represents in situ polymerization of monomer reactants.

**polar winding.** A winding in which the filament path passes tangent to the polar opening at one end of the chamber and tangent to the opposite side of the polar opening at the other end. Also known as planar winding.

**polyesters.** Thermosetting resins, produced by dissolving unsaturated, generally linear alkyd resins in a vinyl-type active monomer such as styrene, methyl styrene, or diallyl phthalate. Cure is effected through vinyl polymerization using peroxide catalysts and promoters, or heat, to accelerate the reaction. The resins are usually furnished in solution form, but powdered solids are also available.

**polyetheretherketone (PEEK).** A linear aromatic crystalline thermoplastic. A composite with a PEEK matrix may have a continuous-use temperature as high as 480 °F (250 °C).

**polymer.** A high-molecular-weight organic compound, natural or synthetic, whose structure can be represented by a repeated small unit, the mer, for example, polyethylene, rubber, and cellulose. Synthetic polymers are formed by addition or condensation polymerization of monomers. Some polymers are elastomers, some are plastics. When two or more monomers are involved, the product is called a copolymer.

**postcure.** Additional elevated-temperature cure, usually without pressure, to improve final properties and/or complete the cure, or decrease the percentage of volatiles in the compound. In certain resins, complete cure and ultimate mechanical properties are attained only by exposure of the cured resin to higher temperatures than those of curing.

**pot life.** The length of time that a catalyzed thermosetting resin system retains a viscosity low enough to be used in processing. Also called working life.

**preimpregnation.** The practice of mixing resins and reinforcement and effecting partial cure before use or shipment to the user. See also *prepreg*.

**prepreg.** Ready-to-wind material in roving form that is impregnated with resin and stored for use. The resin generally is partially cured to a B-stage and supplied to the fabricator, who winds the finished shape and completes the cure with heat. A wet prepreg contains resin that is not preadvanced and is subject to the same viscosity changes as a wet-wound resin system.

## R

**reinforced plastics.** Molded, formed, filament-wound, tape-wrapped, or shaped plastic parts consisting of resins to which reinforcing fibers, mats, fabrics, and so forth have been added before the forming operation to provide some strength properties greatly superior to those of the base resin.

**reinforcement.** A strong material bonded into a matrix to improve its mechanical properties. Reinforcements are usually long fibers, chopped fibers, whiskers, particulates, and so forth. The term should not be used synonymously with *filler*.

**release agent.** A material that is applied in a thin film to the surface of a mold to keep the resin from bonding to the mandrel.

**resin.** A solid, semisolid, or pseudosolid organic material that has a variable (often high) molecular weight, exhibits a tendency to flow when subjected to stress, usually has a softening or melting range, and usually fractures conchoidally. Most resins are polymers. In reinforced plastics, the material used to bind together the reinforcement material; the matrix. See the Appendix for specifications and designations of resins.

**roving.** A collection of bundles of continuous filaments either as untwisted strands or as twisted yarns. Rovings may be lightly twisted, but they generally are wound as bands or tapes with as little twist as possible.

**roving ball.** A term used to describe the supply package offered to the winder. It consists of a number of ends or strands wound to a given outside diameter onto a length of cardboard tube.

## S

**S-glass.** A magnesium aluminosilicate glass specially designed to provide very high-tensile-strength glass filaments. S-glass and S-2 glass fibers have the same glass composition but different finishes (coatings). S-glass is made to more demanding specifications, and S-2 is considered the commercial grade.

**single-circuit winding.** A winding in which the filament path makes a complete traverse of the component, after which the following traverse lies immediately adjacent to the previous one. See also *polar winding.*

**size.** A treatment applied to fiber to protect its surface.

**skein.** A continuous filament, strand, yard, roving, and so on wound up to some measurable length.

**specific properties.** Material properties divided by material density.

**spun roving.** A heavy, low-cost glass fiber strand consisting of filaments that are continuous but doubled back on each other.

**stacking sequence.** A description of a laminate that details the ply orientations and their sequence in the laminate. Also known as winding sequence.

**stiffness.** The relationship of load and deformation; a term often used when the relationship of stress to strain does not conform to the definition of Young's modulus.

**strain gage.** Device to measure strain in a stressed material based on the change in electrical resistance.

**strand.** An assembly of continuous filaments, without twist. When used as a unit, may include tows, ends, yarn, and denier.

**stress.** The internal force per unit area that resists a change in size or shape of a body. Expressed in force per unit area.

## T

**tangent line.** The axial distance from the outside of the pressure vessel at the nearest dome to the intersection of the dome and the cylinder.

**tenacity.** The term generally used in yarn manufacture and textile engineering to denote the strength of a yarn or of a filament of a given size. Numerically, it is the grams of breaking force per denier unit of yarn or filament size (grams per denier, or gpd). The yarn is usually pulled at the rate of 12 in./min. Tenacity equals breaking strength (grams) divided by denier.

**tensile strength.** The maximum load or force per unit cross-sectional area, within the gage length, of the specimen. The pulling stress required to break a given specimen.

**tex.** A metric term used to classify glass; it refers to grams per 1000 m of material.

**thermoplastic.** Capable of being repeatedly softened by an increase in temperature and hardened by a decrease in temperature. Applicable to those materials whose change upon heating is substantially physical rather than chemical and that, in the softened stage, can be shaped by flow into articles by molding or extrusion.

**thermoset.** A plastic that, when cured by application of heat or chemical means, changes into a substantially infusible and insoluble material.

**torsion.** Twisting stress.

**toughness.** A property of a material for absorbing work. The actual work per unit volume or unit mass of material that is required to rupture it. Toughness is proportional to the area under the load-elongation curve from the origin to the breaking point.

**tow.** An untwisted bundle of continuous filaments. Commonly used in referring to man-made fibers, particularly carbon and graphite but also glass and aramid. A tow designated as 140 has 140,000 filaments.

**towpreg.** Preimpregnated tow or strands for filament winding. It differs from typical prepreg tape in that the tow is wound onto the roving core without a separator sheet. The tow matrix has been adjusted in tack so that it unwinds freely. Also, the tow can have more lateral displacement on a mandrel than an equivalent-width tape because the individual tows can slip relative to the rest of the band.

**turns per inch (tpi).** A measure of the amount of twist produced in a yarn, tow, or roving during its processing history. Also, the lead rate of a hoop layer at a specified band width.

## V

**virgin filament.** An individual filament that has not been in contact with any other fiber or any other hard material.

**viscosity.** The property of resistance to flow exhibited within the body of a material, expressed in terms of relationship between applied shearing stress and resulting rate of strain in shear. Viscosity is usually taken to mean Newtonian viscosity, in which case the ratio of shearing stress to the rate of shearing strain is constant. In non-Newtonian behavior, which is the usual case with plastics, the ratio varies with the shearing stress. Such ratios are often called the apparent viscosities at the corresponding shearing stresses.

**voids.** Air or gas that has been trapped and cured into a laminate. Porosity is an aggregation of microvoids. Voids are essentially incapable of transmitting structural stresses.

**volatiles.** Materials in a sizing or a resin formulation that are capable of being driven off as a vapor at room, or slightly elevated, temperature.

## W

**wafer.** A reinforcement for port openings.

**wet-out.** The condition of an impregnated roving or yarn wherein substantially all voids between sized strands and filaments are filled with resin.

**wet winding.** A process in which the fiber reinforcement is coated with resin just prior to wrapping on the mandrel.

**wind angle.** The angular measure in degrees between the direction parallel to the filaments and an established reference. In filament-wound structures, it is the convention to measure the wind angle with reference to the centerline through the polar bosses (that is, the axis of rotation).

**winding pattern.** A regularly recurring pattern of the filament path after a certain number of mandrel revolutions, leading to the eventual complete coverage of the mandrel.

**winding tension.** In filament winding, the amount of tension on the reinforcement as it makes contact with the mandrel. Winding tension can influence resin content, propensity to slip at a specific wind angle, and residual stress, among others.

## Y

**yarn.** An assemblage of twisted filaments, fibers, or strands, either natural or manufactured, to form a continuous length that is suitable for use in weaving or interweaving into textile materials.

**yield.** An English term used to classify glass; it refers to nominal yards of material per pound.

**Young's modulus.** The ratio of normal stress to corresponding strain for tensile or compressive stresses less than the proportional limit of the material.

## Z

**zero-degree winds.** Specialty winds for generating zero-angle winds that run the length of cylinders (generally combined with the use of specialized mandrels to catch the fiber at the ends of the part and prevent slippage).

The following sources were used in this compilation:


- *Composites*, Vol 1, *Engineered Materials Handbook*, ASM International, 1987
- "Glossary of Filament Winding Terms," McClean Anderson, http://www.mccleananderson.com/support/glossary.html
- C.D. Hermansen and R.R. Roser, "Filament Winding Machine: Which Type Is Best for Your Application?" 36th Annual Conference, Reinforced Plastics/Composites Institute, The Society of the Plastics Industry, Feb 1981
- G. Lubin, Ed., *Handbook of Composites*, Van Nostrand Reinhold Co., New York, 1982
- S.T. Peters, W.D. Humphrey, and R.F. Foral, *Filament Winding: Composite Structure Fabrication*, 2nd ed., SAMPE Publishers, Covina, CA (© S.T. Peters, 2009)
- A.M. Shibley, H.L. Peritt, and M. Eig, "A Survey of Filament Winding: Materials, Design Criteria, Military Applications," Plastic Report 10, Plastics Technical Evaluation Center, U.S. Armament Research and Development Command, Dover, 1962

Copyright © 2011, ASM International®
All rights reserved.
www.asminternational.org

# Index

## C



## D

## X



## Y



## Z

FSC
www.fsc.org
MIX
Papier | Fördert
gute Waldnutzung
FSC® C083411

Kolibri 360 GmbH
Ferdinand-Jühlke-Straße 7
99095 Erfurt, Deutschland
produktsicherheit@kolibri360.de